S0-CAH-608

Plants in Saline Environments

Edited by

A. Poljakoff-Mayber and J. Gale

Contributors

D. L. Carter   V. J. Chapman   L. D. Doneen

J. Gale   A. Kylin   A. J. Peck   A. Poljakoff-Mayber

R. S. Quatrano   I. Shainberg   W. W. Thomson

With 54 Figures

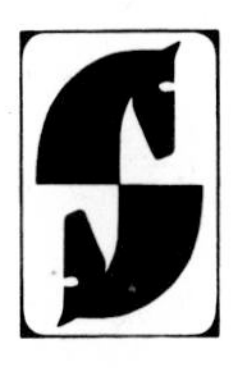

Springer-Verlag New York Heidelberg Berlin 1975

The picture on the cover is taken from Chapter 1, p. 11.

ISBN 0-387-07193-8 Springer-Verlag New York · Heidelberg · Berlin
ISBN 3-540-07193-8 Springer-Verlag Berlin · Heidelberg · New York

Distributed in the British Commonwealth Market by Chapman & Hall Limited, London.

Library of Congress Cataloging in Publication Data. Poljakoff-Mayber, Alexandra, 1915– Plants in saline environments. (Ecological studies; v. 15) Bibliography: p. . Includes index. 1. Halophytes. 2. Plants, Effect of salts on. 3. Soils, Salts in. 4. Saline irrigation. 5. Salinity. I. Gale, Joseph, 1931– joint author. II. Title. III. Series.
QK922.P65 581.5'26 75-1272

Typesetting, printing, and binding: Brühlsche Universitätsdruckerei, Gießen.

Contents

Contributors

CARTER, D. L.	Snake River Conservation Research Center, Kimberly, Idaho, USA
CHAPMAN, V. J.	Department of Botany, University of Oakland, Private Bag, Auckland, New Zealand
DONEEN, L. D.	Department of Water Science and Engineering, University of California, Davis, California, USA
GALE, J.	Department of Botany, The Hebrew University of Jerusalem, Jerusalem, Israel
KYLIN, A.	Botanical Institute, University of Stockholm, S 104-05 Stockholm, Sweden. (New permanent address: Department of Plant Physiology, Royal Veterinary and Agricultural University, DK-1871 Copenhagen, Denmark)
PECK, A. J.	CSIRO, Division of Land Resources Management, Private Bag, P.O. Wembley, 6014 W. Australia
POLJAKOFF-MAYBER, ALEXANDRA	Department of Botany, The Hebrew University of Jerusalem, Jerusalem, Israel
QUATRANO, R. S.	Department of Botany and Plant Pathology, Oregon State University, Corvallis, Oregon, 97331, USA
SHAINBERG, I.	Institute of Soils and Water, The Volcani Center, Agricultural Research Organisation, Bet Dagan, Israel
THOMSON, W. W.	Department of Biology, University of California, Riverside, California, USA

Introduction

A. POLJAKOFF-MAYBER and J. GALE

The response of plants to saline environments is of interest to people of many disciplines. In agriculture the problem of salinity becomes more severe every year as the non-saline soils and the non-saline waters become more intensively and more extensively exploited. Further expansion of agriculture must consider the cultivation of saline soils and the use of water with a relatively high content of soluble salts. Moreover, industrial development in many countries is causing severe water pollution, especially of rivers, and mismanagement in agriculture often induces secondary salinization of soils and sources of irrigation water. From the point of view of agriculture it is, therefore, of the utmost importance to know the various responses of plants to salinity and to understand the nature of the damage caused by salinity to agricultural crops.

Botanists and plant physiologists study plants, their form, growth, metabolism and response to external stimuli. A challenging problem for them is to understand the differences between glycophytes, plants growing in a non-saline environment and halophytes, plants which normally grow in salt marshes, in sea water or in saline soils. This includes the elucidation of structural and functional adaptations which enable halophytes to tolerate the saline environment, and also questions as to whether they only tolerate the saline environment or actually thrive in it.

Ecologists and environmentalists are interested in the interrelationships between the organism, in this case the plant, and its environment, from the climatic, edaphic and biotic points of view.

Numerous reviews have been published during recent years dealing with the effects of salinity in agriculture and on various aspects of plant life. BERNSTEIN and HAYWARD (1958) discussed the varying tolerance of different plant species and the effect of salt on the water balance of the plant. STROGONOV's book (published in the Soviet Union in 1962), published in English in 1964, describes various responses of glycophytes and halophytes, emphasising the different effects of various types of salinity. STROGONOV and KABAROV (1964), discussed changes in metabolic pathways and the production of toxic intermediates under various types of salinity. BOYKO (1966) argued the possibility of utilizing saline water for irrigation and discussed the concept that a balance between different species of ions may be less toxic than equiosmolar concentrations of single salts. JENNINGS (1968) suggested an overall theory of salt damage based on the expenditure of energy due to ATPase activation by salt. STROGONOV et al. (1970) summarize the research work done in the Soviet Union on structure and function of plant cells under saline conditions. This summary to some extent overlaps his previous review (1964). Chapters on physiological responses of plants to salinity are included in a recent book of LEVITT (1971). TALSMA and PHILIP (1971) summarize the discussions of a symposium on salinity and water use. WAISEL (1972) devoted his monograph to the biology of halophytes. GREENWAY (1973), in a short but comprehensive review of plant response to salinity, suggested that energy expenditure, during osmotic adjustment to salinity, is one of the main factors in reducing growth. MEIRI and SHALHEVET (1973) in Volume 5 of this series discuss aspects of crop growth under saline conditions, while physical aspects of soil water and salts in ecosystems are reported in Volume 4 (HADAS et al., 1973).

This book attempts to open up new vistas to the scientist proficient in perhaps only one aspect of the salinity problem, and should give the interested reader an overall view of the different problems raised by salinity.

The first Chapter (The Salinity Problem in General, Its Importance and Distribution) by CHAPMAN, gives a general survey of the problem with special reference to halophytes. It reviews the distribution of saline areas and halophytic vegetation throughout the world, and discusses the economic importance and possible uses of saline soils and of halophytic species.

The second Chapter (Problems of Salinity in Agriculture) by CARTER, emphasizes the severity of the salinity problem in modern agriculture and points to some possible practices which will enable the farmer to cope with and combat the damage caused by salinity.

The third Chapter by SHEINBERG (Salinity of Soils — Effects of Salinity on the Physics and Chemistry of Soils), deals with soil salinity, and the fourth Chapter by DONEEN (Water Quality for Irrigated Agriculture), deals with problems of salinity in water (especially in irrigation water). Both writers cover their subjects mainly from the agricultural point of view. They clarify and explain the concepts used in these fields and discuss the practices used in agricultural management of saline soils and waters.

The fifth Chapter by PECK (Effects of Land Use on Salt Distribution in the Soil) deals with the factors influencing distribution of soluble salts in the natural environment, and the redistribution which results from environmental disturbances, such as the cutting down of forests for the creation of pasture.

The characteristics of the soil and water are the factors which determine whether or not the environment will be saline. Soil and water also constitute the main source of nutritive mineral ions for plants. Mineral nutrition of plants in general and ion uptake and transport in relation to salinity in particular, were very recently reviewed by EPSTEIN (1972) and by RAINS (1972). Consequently, these aspects of the salinity problem are discussed only very briefly in the intermediary remarks between Chapters 5 and 6.

Chapters 6–10 deal with plants and their various responses to saline environments. When reviewing and discussing the behavior of plants under saline conditions the main problem is how to deal with the multitude of different responses which have been reported. Changes in plant behavior induced by salinity have been found in water uptake and water balance, gas exchange—transpiration, photosynthesis and respiration, optical properties of leaves, ion uptake, metabolic pathways, growth, morphology and anatomy of the plant, and balance of hormones. Furthermore, the changes induced by salinity in any one of the particular physiological or anatomical parameters varies considerably depending on the plant species, the stage of its development and external factors such as edaphic conditions, species of salt involved, salt regime and climatic conditions such as heat and humidity. Consequently it is very difficult to quantitize plant responses to salinity in a way which could be meaningful for extrapolation from species to species, or from one set of environmental conditions to another.

The general effects of salinity on growth and structure are discussed in Chapter 6 by POLJAKOFF-MAYBER (Morphological and Anatomical Changes in Plants in Response to Salinity). The seventh Chapter by THOMSON (The Structure and Function of Salt Glands) is devoted to salt glands, the special organs typical of salt excreting halophytes. Metabolic and biochemical aspects of salt tolerance, mainly on the cellular level, are discussed in Chapter 8 by KYLIN and QUATRANO (Metabolic and biochemical aspects of salt tolerance), while in Chapters 9 (Plant Water Balance and Gas Exchange under Saline Conditions) and in 10 (The Combined Effect of Environmental Factors and Salinity on Plant Growth), GALE discusses physiological aspects of salinity at the whole plant level. Finally, in the general discussion, the editors attempt to evaluate the different aspects of the salinity problem and to present an overall picture of how plants respond to the stress imposed on them by saline environments, whether natural or man made.

References

BERNSTEIN, L., HAYWARD, H. E.: Physiology of salt tolerance. Ann. Rev. Plant Physiol. **9**, 25–46 (1958).

BOYKO, H. (Ed.): Salinity and aridity. The Hague: Junk Publ. 1966.

EPSTEIN, E.: Mineral nutrition of plants: principles and perspectives. New York: John Wiley and Sons 1972.

GREENWAY, H.: Salinity, plant growth and metabolism. J. Aust. Inst. Agr. Sc. Vol. 1973, March., 24–34 (1973).

HADAS, A., SWARTZENDRUBER, D., RIJTEMA, P. E., FUCHS, M., YARON, B. (Eds.): Physical aspects of soil water and salts in ecosystems. Ecological Studies, Vol. 4. Berlin-Heidelberg-New York: Springer 1973.

JENNINGS, D. H.: Halophytes, succulence, and sodium in plants, a unified theory. New Phytol. **67**, 899–911 (1968).

LEVITT, J.: Response of plants to environmental stresses. New York: Academic Press 1972.

MEIRI, A., SHALHEVET, J.: Crop growth under saline conditions. In: YARON, B., DANFORS, E., VAADIA, Y. (Eds.): Ecological Studies, Vol. 5, pp. 277–290. Berlin-Heidelberg-New York: Springer 1973.

RAINS, D. W.: Salt transport by plants in relation to salinity. Ann. Rev. Plant. Physiol. **23**, 367–388 (1972).

STROGONOV, B. P.: Physiological basis of salt tolerance of plants. Jerusalem: I. P. S. T. 1964.

STROGONOV, B. P., KABANOV, V. V.: Salt tolerance of plants. In: KOVDA, V., HAGAN, R. M., VAN DEN BERG, C. (Eds.): International sourcebook on irrigation and drainage in relation to salinity and alkalinity. Rome: FAO/UNESCO 1964.

STROGONOV, B. P., KABANOV, V. V., SHEVJAKOVA, N. I., PALINA, L. P., KOMIZERKO, E. I., POPOV, B. A., DOSTANOVA, R. KH., PRYKHOD'KO, L. S.: Struktura i Funktziya Kletok rastenii pri Zusolenii (Structure and function of plant cells under salinity). Moskow: Nauka 1970.

TALSMA, T., PHILIP, J. R.: Salinity and water use. London: MacMillan 1971.

WAISEL, Y.: Biology of halophytes. New York: Academic Press 1972.

General Review of the Salinity Problem

CHAPTER I

The Salinity Problem in General, Its Importance, and Distribution with Special Reference to Natural Halophytes

V. J. CHAPMAN

A. Introduction

A salinity problem is regarded as arising when the concentration of sodium chloride, sodium carbonate, sodium sulphate or salts of magnesium are present in excess, and when the effect becomes increasingly evident the greater the excess. There has, however, been considerable debate over the exact point, in respect of salt concentration, where the salinity problem first appears, i.e. the point where one passes from glycophytic to halophytic conditions and vice versa. At present only sodium chloride of the above alkali salts has really been studied sufficiently, and there is no evidence to suggest any departure from the earlier figure of 0.5% in the soil solution (CHAPMAN, 1966; CHAPMAN, 1974). It is true that there are some economic plants that can be grown satisfactorily using irrigation water containing 1% NaCl provided the substrate has a high permeability, e.g. sand (BOYKO, 1966). It is also evident that the effect of increased salinity upon plants is determined not only by the absolute quantity of the ions in excess, but also by the relative amounts of certain other ions, especially $SO_4^=$ (STROGONOV, 1964). Some excess is not necessarily deleterious, because BOYKO (1966) has suggested that with certain plants, temporary or permanent salinization may give increased resistance to drought and disease, and this led him to propound his concept of "raised vitality".

The halophytic habitat is automatically associated with the special types of plants, halophytes, that are capable of growing under such conditions. In spite of the considerable interest displayed in these plants and the amount of experimental work performed on them, there is still some doubt about the number of species that are obligate halophytes[1] (BARBOUR, 1970). Some such genera undoubtedly exist, e.g. *Salicornia*, *Rhizophora*, *Zostera*, but it does seem that most of the species occupying saline habitats are facultative halophytes and have been restricted to the habitat as a result of failure to compete under glycophytic conditions. It is evident that this is an aspect that would still repay further investigation.

The saline areas of the world comprise the salt marshes of temperate latitudes, the mangrove swamps of the sub-tropics and tropics, interior salt marshes found adjacent to salt lakes, e.g. the Great Salt Lake in the USA, Neusiedler See in Austria, salt deserts and smaller areas around salt springs.

[1] Species that reach their optimum growth at salinities in excess of 0.5% NaCl.

B. Salinity Problems

Problems posed by excess of salinity have till now been primarily of academic interest. The growing world population is however forcing this academic interest to become an important economic one because saline lands, especially maritime areas, are rich in other salts, some of them valuable trace elements required in plant growth. If the excess salt can be removed such lands are very valuable agriculturally. On the other hand we cannot regard with equanimity the loss of all, or even a major proportion, of estuaries, marshes and mangrove swamps. In recent years ecologists have emphasised that these wetlands support vast numbers of fish and birds and this makes them one of the world's most productive ecosystems. One of the most pressing problems, therefore, is an economic appraisal of the total value of saline areas under wild conditions and after reclamation, so that some firm recommendations can be made as to how much can safely be reclaimed without causing damage to the overall environment.

It appears that at present we possess a substantial body of information about the factors responsible for excess of salinity in many parts of the world. Such factors include frequency of tidal inundation, precipitation, effect of drainage channels, soil characteristics, vegetation communities, water table, depth of salt deposits and temperature. Water inflow and outflow from any saline area is obviously of great significance, but whilst this is appreciated it does seem that at present we do not possess sufficient factual data. This, then is a field worthy of further investigation.

Some specific habitats have received very much more attention than others. Temperate salt marshes, inland salt deserts and estuarine mangrove swamps are fairly well understood. Lagoon mangroves, which are frequent in the Gulf of Mexico (THOM, 1967) and part of West Africa, are worthy of further study, so also are the mangrove forests of Western South America, because they contain some species, e.g. *Pelliciera rhizophorae*, not found elsewhere (WEST, 1963).

The effect of salinity upon morphological features of plants has been documented, but this is an area which requires further study. STROGONOV (1964) has reported different effects upon different species, and if we are to understand these responses many more studies will be necessary. So far only one major study has been reported making use of computer techniques (WEIHE, 1963). The computer makes it possible to analyze the inter-relations between a wide range of habitat factors and morphological features, and more such studies are highly desirable. The increasing awareness of the effect of different ions upon features such as succulence, pubescence, leaf reduction etc. indicates that a wider range of studies would be of value. An insufficient number of wild species as well as of cultivated plants have so far been investigated, and whilst further studies may only serve to confirm what has already been established the extra data will be invaluable.

CHAPMAN (1966) has pointed out that published results indicate a variation in tolerance towards salinity by both wild and cultivated species and that this variation seems to depend upon the locality. Further studies should be prosecuted in order to establish whether it is differences in climate, soil or in genetic make-up of the species that is primarily responsible. A further examination of the possibility of physiological races is also desirable. Whilst SCHOLANDER (1968) and his co-workers have made an attack upon the problem of exudation of salt from mangrove species, much less work has been carried out on salt marsh and salt desert herbs. The most recent is that of POLLAK and WAISEL (1970) on salt excretion in *Aeluropus*. Salt excretion is closely allied to the problems of protoplasmic permeability to the various saline ions (Na^+, Mg^{2+}, Cl^-, $CO_3^=$) and our present knowledge concerning the uptake, transport and excretion of these is still inadequate. Whilst we possess some information about the inter-action between ions with respect to plant uptake, this is primarily concerned with potassium and sodium, and from the point of view of soil reclamation and species suitable for planting we ought to know more about other ions in relation to sodium, magnesium and chloride.

One field area that has been grossly neglected is that of the microorganisms present in saline soils. Until detailed studies of these, such as that by KELIER and HENIS (1970), are carried out, we shall not have a full understanding of the rhizosphere nor of the biological processes taking place in the soil. Studies of this nature are also basic to an understanding of what may happen when such soils are subject to reclamation processes.

It has been known for a long time that certain plant species can be used as indicators of degrees of soil salinity. HABIB et al. (1971) have reported on the native indicators of soil salinity in Irak. *Aster subulatus* in the USA and as an introduced species in New Zealand, is a recognised indicator of brackish conditions. However, no really detailed studies have been carried out on a range of native species in different parts of the world in order to establish just what range of salinity values they can indicate. This would seem to be an important exercise from the aspect of potential reclamation.

The wide-spread development of aerenchyma tissue in many salt marsh and mangrove species indicates that gaseous diffusion represents a real problem, especially in maritime soils where the sodium ion exerts an unfavorable influence upon the soil colloids. It has been clearly established that gaseous exchange and internal diffusion takes place in both mangrove (CHAPMAN, 1944) and salt marsh plants (TEAL and KANWISHER, 1966; 1970) but the number of species investigated is a mere fraction of those known to possess aerenchyma. The mere establishment of internal gaseous diffusion is one aspect whilst the effect of salinity variations upon such diffusion is another.

The presence of excess of salt in the soil environment presents problems of its control. One possibility is the actual exclusion of salt by the roots. This appears to be the case in certain mangrove genera (SCHOLANDER, 1968). Another possibility is the development of water storage tissue in order to reduce the otherwise high osmotic pressures that would develop. This type of control is exemplified by certain mangroves and also by *Salicornia* and *Suaeda* among other genera. A third type of control is represented by the annual loss of organs with their contained salt. Some perennial succulents behave in this fashion, as also does *Juncus maritimus*. Most of the work in this area has been carried out upon maritime halophytes and very few studies have been made upon salt desert plants.

Whilst there may be many other problems associated with saline habitats, the final one to be considered here is that of seed germination. It is true that a number of studies have been executed in this field and it is evident that germination, even of obligate halophytes, only takes place under conditions of reduced salinity. We do possess some information about the seasonal reduction of salinity on maritime salt marshes which enables seed germination to take place, but there seems to be little information available about seed germination, or the conditions necessary to secure it, in salt deserts. Salinity, of course, may not be the only factor controlling germination of seeds in saline areas. There is published evidence to show that, for some species at least, either temperature or light may be an additional factor (see CHAPMAN, 1974). It is evident that there is an area here that provides plenty of scope for further studies.

C. Distribution of Saline Areas

Maritime salt marshes, mangrove swamps and interior salt deserts are all available for reclamation, and, in view of the increasing demand of the nations for more food, it is likely that more and more of this type of land will be reclaimed for agricultural purposes. In parts of the world, e.g. Europe, USA, Malaya, some of the land has already been reclaimed, and in other parts, mangrove forest is managed as a source of pulp (Bangladesh), cutch (E. Africa and Indonesia), and charcoal (Malaysia) as well as providing local building timber.

The soil of salt marshes and mangrove swamps is generally rich in nutrients, much of it being virgin soil produced by the action of rivers or sea erosion. It is a well-established fact that estuaries are noted for their high biological activity associated with the mixing of water that occurs there. There is a rich supply of nutrients brought up from the deeper sea waters, a steady supply of nutrients from land water runoff and, in places, further enrichment from the effluent of human settlements and cities (ANON, 1972).

About 70% of the earth's population live within an easy day's travel of the coast and man's needs have resulted in the development of many ports and associated industrialization. All this has led to extensive misuse of the highly productive salt marshes and mangrove swamps. Land that should have been reclaimed for agricultural purposes has been reclaimed for settlement or industry. Between 1922 and 1954 one quarter of the USA salt marshes were destroyed in this manner (TEAL and TEAL, 1969). Between 1954 and 1964 a further 10% of the salt marsh between Maine and Delaware was destroyed. No figures appear to be available for similar destruction in Europe. In Australia and New Zealand hungry, expanding cities see mangrove swamps and salt marshes as land ready for reclamation for industrial purposes or as suitable sites for refuse dumps. An awakened population, e.g. in Auckland, is protesting strongly about the latter use because of its effect upon the environment.

It is difficult to determine the extent of the inland saline and alkali lands because there is no accepted criterion as to when a soil is to be regarded as belonging to one category or the other.

In most cases salinity of inland areas is related to high aridity plus a saline water table from rocks rich in sodium salts. There are low physiographic gradients so that the water accumulates rather than draining away (HAYWARD, 1954).

The following summary is compiled from various sources but much information has been derived from RAHEJA (1966) (Fig. 1).

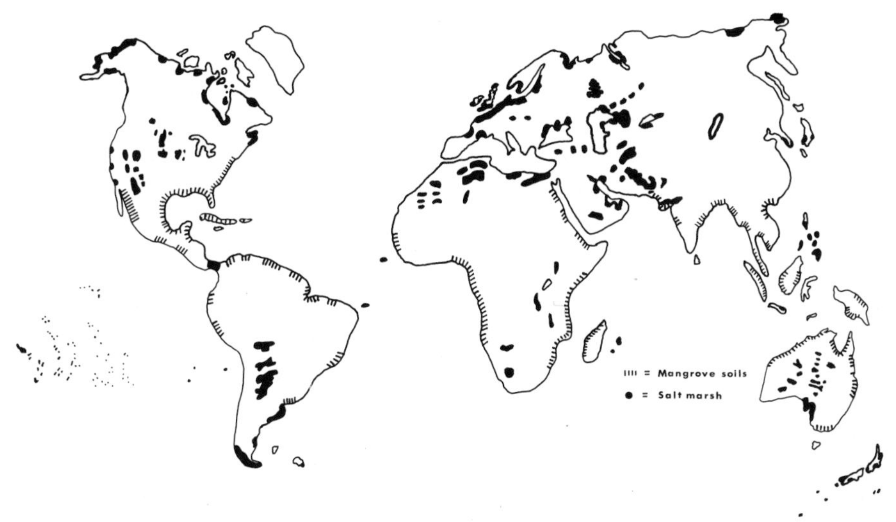

Fig. 1. Distribution of saline soils

I. Australia

Saline soils in Australia are mainly located in the southern part of the continent, especially where the annual rainfall is less than 42 cms. Extensive maritime salt soils occur all through the north and north east and again in the south between Sydney and Adelaide. In West and South Australia, Victoria and New South Wales, large areas are occupied by salinized brown soils, known as "mallee" soils, in which the exchangeable sodium has been built up by leaching. These soils cover about 55% of the total area. Solonetz soils[2], covering 0.6% of the total area are to be found in West and South Australia and in Queensland (e.g. lower Burdekin valley). It is to be noted that some of the Queensland soils have been produced as a result of secondary salinization.

II. Indian Sub-Continent

In India and Pakistan there are about 20 million acres of saline or salinized soils quite independent of all those maritime soils associated with coastal mangroves. Many of these soils are termed "reh" soils and contain salts of sodium brought to the surface by capillary action and evaporation. Saline mangrove soils are abundant around Karachi, Madras, the Godavari and Krishna deltas, and the deltas of the Ganges and Brahmaputra. In the Punjab area there are some 5 million acres of soil in which the predominant salts are sodium carbonate and bicarbonate and sodium sulphate. Some 3 million acres ($^3/_5$ of the total) have become salinized as a result of irrigation and by 1960 1.3 million acres had gone out of productivity because the salt concentrations were too high. In that year ASGHAR

[2] Saline soil with a definite structure and sodium commonly present as carbonate, sometimes with sulphate and chloride.

(1960) reported that 100000 acres a year were steadily becoming salinized. In Uttar Pradesh saline soils with excess sodium carbonate and chloride are to be found and there has been some salinization as a result of irrigation. In Maharashtra the main salts are sulphate and chloride and there is less evidence of salinization due to irrigation. Around the Rann of Cutch in Gujarat there are 800 sq. miles of saline soils.

III. Africa

In Africa there are vast saline areas along the north coast with extensions into Eastern Mediterranean countries. In Egypt some 300000 acres have become saline as a result of irrigation from the Nile and extensive efforts are being made to reclaim them. In both Morocco and Israel major efforts are being made to reclaim saline areas and use them for crops. Similar areas exist in central, south east and south west Africa but their extent is unknown, and very little investigational work appears to have been carried out upon them. In Somaliland also there are major saline areas but the arid climate would inhibit their reclamation without extensive irrigation. Wide-spread mangrove soils are found on the west coast between Gabon and Angola and on the east coast between Lamu and Natal. Similar extensive areas are also found on the east coast of Madagascar.

IV. Europe

Very considerable areas of saline soils exist in Europe. The exact extent of the salt marshes of the English Channel and North Sea is not fully documented, whilst some at least of the inland areas have now been reclaimed. Recent figures suggest 40000 acres each in England and Scotland respectively[3]. In Rumania there was originally some 300000 hectares but this is now reduced (OPREA, 1965). In Austria there is a considerable area of wild marsh around the Neusiedler See and in Hungary there are some 500000 hectares, mostly in the Great Hungarian Plain, where the seasonally fluctuating water table favors salinization. There is also an extensive area in Bulgaria (OVDENICHAROV, 1961).

V. USSR

In the USSR saline and alkali soils occupy about 3.4% of the land area (=75 million hectares) and only a small percentage has so far been put to any use. Solontchak soils[4] predominate in Uzbekistan, Tadjikistan, Eastern Kazakhstan, Azerbaijan, Georgia and Turkmenistan, whereas solonetz soils predominate in Western Siberia and Western Kazakhstan. These large inland deserts primarily include basins contained within mountain ranges, e.g. the Solontchak deserts of central Iran, the Western Siberian depression between the Altai and Ural mountains (ORLOVSKII, 1965) and the Turkestan depression. Most of these are charac-

[3] G. COLE, pers. comm.
[4] Soil without definite structure and sodium chloride the principal salt.

terized by a high sodium chloride content with or without excess sodium sulphate and magnesium. There are also inadequately drained alluvial plains such as those of the Amur-Daria, Syr-Daria, the lower Mesopotamian basin (BOUMANS and HUSBOS, 1960), the Kura-Araks lowlands (MURATOVA, 1961), the Karabah plain (AGAEV, 1965) and the valley of the Vakch, the soils of which are rich in chloride and sulphate. Soils with secondary salinization arising from irrigation are to be found in the Vakch valley, Golodhaya Steppe (now being steadily reclaimed) and the Ferghana and Boukhara oases.

VI. North America

In the New World in Canada saline soils exist in all the seaboards, including the Arctic zone, but in addition there are interior saline soils, mainly restricted to the prairie regions of Alberta and Saskatchewan. Here the salts are commonly sulphates of magnesium and sodium, especially the latter (LÜKEN, 1961). CAIRNS (1969) has estimated that there are between 8–12 million hectares of such soils with "low productivity, low infiltration rate, poor root penetration and a narrow moisture profile below the A horizon"[5].

In the USA the salt problem is mainly associated with the Great Salt Lake basin, the interior valleys of California (San Joachim, Sacramento, Caochella, Imperial), the Colorado and Rio Grande drainage basins and parts of the Columbia and Missouri river basins. In the south-western region it has been estimated that 300000 acres have been salinized on account of non-provision of drainage. Extensive maritime salt-soils exist along the entire Atlantic seaboard from the Bay of Fundy down to Florida and this continues along the coastline of the Gulf of Mexico. Much smaller areas are found on the Pacific coast where physiography has not favored their development.

VII. Central and South America

In Central America saline soils, with about 90% exchangeable sodium in the base exchange complex, occur down the west coast and in the inland arid region of the west coast. Soils from the dried out Lake Texcoco also have a high percentage of salt in them. Irrigated saline soils occur in the Mexicale district, along the Rio Grande south of El Paso, and in the lower reaches of the river, including almost all soils of the flood plain. Extensive mangrove soils are to be found on the eastern coast.

In South America large areas of mangrove swamp exist all along the northern coast, especially in lagoons, and then south on the east coast, the greatest areas being at the mouth of the Amazon. On the west coast mangrove soils are less extensive and do not occur south of 4° latitude. In the southern area salt marsh soils are to be found but their extent is quite unknown. In the highly arid Pacific areas of S. Equador, Peru, Argentine and Chile there are large inland areas of

[5] Soil horizon characterized by major organic matter accumulation and loss of clay, iron and aluminum by leaching.

saline soils. The Chilean soils are essentially natural, as are those of the Salinas Grandes in the Argentine, but in Peru and Equador the soils have become saline as a result of irrigation in basins with no outlets. In Northern Brazil it is reported that there are about 25000 hectares of saline soil arising from irrigation with brackish water.

D. Distribution of Maritime Salt Marsh Vegetation

Maritime salt marshes are widespread, especially in temperate parts of the world (Fig. 1). The dominant vegetation is essentially phanerogamic herbs, though some shrubs do occur. There may also be an extensive algal vegetation. In general, marshes in specific regions tend to be characterized by typical groups of plants. This fact can be used to classify the marshes into distinct areas. Based on this premise the major groups of maritime salt marshes are as follows:

I. Arctic Group

These marshes extend around the arctic circle and exist under an extreme environment so that species are few and the succession is simple. The dominant species is the grass *Puccinellia phryganodes* with two species of *Carex*, *C. subspathacea*, and *C. maritima* being important at upper levels.

II. North European Group

This includes marshes from Spain northwards around the English Channel and all North Sea and Baltic Coasts as well as around the shores of Gt. Britain and Eire. Sub-groups can be recognized which are related to soil differences, e.g. in S.W. Eire, to salinity differences, e.g. in the brackish Baltic, or to the presence of aggressive species, e.g. *Spartina townsendii* and *S. anglica* in the English Channel. Throughout the region as a whole there is a dominance of annual *Salicornia* species, *Puccinellia maritima*, *Juncus gerardi* and the General Salt Marsh Community.

The habitat of the Scandinavian subgroup is typified by a high proportion of sand in the soil and it has developed on a rising coastline. It is dominated by grasses such as *Puccinellia*, *Festuca rubra* and *Agrostis stolonifera*, and because of this the marshlands are frequently used for grazing by the local farmers. The subgroup can be further subdivided, seven such subgroups being recognized among the British marshes (Chapman, 1974) and Gillner (1960) has presented evidence for the existence of three subgroups in S.W. Sweden.

The North Sea subgroups have much more clay and silt in the soil and there is generally a wider range of plant communities. Grasses are less prominent and herbs of the General Salt Marsh Community are more in evidence. These marshes are associated with a subsiding coastline. At present their character is changing through the introduction or natural spread of the aggressive *Spartina* species. The

British marshes can be subdivided into 5 separate areas (CHAPMAN, 1973b) and the Elbe and Dutch marshes each into 2 separate areas. These minor divisions are based upon a variety of criteria which may be floristic, e.g. Mersea and Suffolk marshes and the Norfolk marshes; algal, e.g. Lincolnshire marshes with no marsh fucoids; salinity, e.g. the normal and hyposaline marshes of the Elbe estuary; and morphogenetic, e.g. the river bank and basin marshes of the Netherlands.

The Baltic subgroup differs from the two preceding groups in the presence of some characteristic species, e.g. *Carex paleacea, Juncus bufonius*, as well as others, e.g. species of *Scirpus*, that play a major part as primary colonists because of the lowered salinity.

The English Channel subgroup was probably originally comparable with the North Sea subgroup but since the appearance of *Spartina townsendii* and *S. anglica* they have a completely different physiognomic appearance. Eventually, it may well be that North Sea marshes will become similar.

III. Mediterranean Group

These marshes carry characteristic species of *Arthrocnemum* and *Limonium*. The western subgroup has clear floristic affinities with the North European group whereas the eastern subgroup has floristic affinities with the interior saline deserts of Eurasia, e.g. the presence of *Halocnemum strobilaceum, Petrosimonia crassifolia, Suaeda altissma*. From the literature it would appear that the Caspian salt marshes represent a third subgroup where species such as *Kalidium caspicum* and *Anabasis aphylla* are to be found.

IV. Western Atlantic Group

These marshes extend from the St. Lawrence to Florida. In the North they pass into the Arctic group and in the South there is a transition to mangrove. They fall into three subgroups, essentially based on method of formation, though there are also floristic differences associated mainly with the southern subgroup. The Bay of Fundy subgroup is formed in front of a weak-rock upland with a thick depth of mud because of the Fundy tides with their great vertical range. The New England subgroup has formed in front of a hard-rock upland and so the soil is a peat from plant remains rather than a muddy clay. The Coastal Plain type is built of a muddy clay because it too has formed in front of a soft-rock upland.

V. Pacific American Group

These marshes extend from Southern Alaska to California with characteristic species not found elsewhere. There does not seem to be any basis for subdividing them.

VI. Sino-Japanese Group

Some of the species, e.g. *Limonium japonicum, Suaeda japonica, Limonium tetragonum, Puccinellia kurilensis* give a distinct character to the group, which in the north begins to exhibit a transition to the Arctic group. Further study in the extensive south will probably show a transition to mangrove swamp.

VII. Australasian Group

These are characterised by some specifically southern hemisphere species such as *Salicornia australis, Samolus repens, Suaeda novae zelandiae, Arthrocnemum* spp. Floristically they evidently form one entity but there are sufficient floristic differences to separate the Australian and New Zealand marshes into two subgroups. The latter lack species such as *Arthrocnemum arbuscula, A. halocnemoides, Limonium australis*, and *Frankenia pauciflora*.

VIII. South American Group

These are found on both coasts of South America, and compared with salt marshes elsewhere are in need of much further study. Floristically there are local species of genera such as *Spartina, Distichlis, Heterostachys* and *Allenrolfea*.

IX. Tropical Group

The marshes generally occur at high levels behind mangrove swamps and are flooded only by extreme tides. Typical species are *Sesuvium portulacastrum* and *Batis maritima*.

E. Distribution of Mangrove Vegetation

In the tropical and subtropical parts of the world maritime salt marshes are replaced by mangrove swamps in which the dominant vegetation are trees, shrubs and a few lianes. In the northern hemisphere mangal[6] extends north to latitudes ranging between 24° and 32° N. Mangrove species occur in Southern Japan, whilst in the Atlantic, Florida and Bermuda represent their northerm limit. On the USA Pacific coast they do not occur north of latitude 24° 38′ N. In the Southern hemisphere mangal is found in Brazil south to the tropic of Capricorn but on the West coast they do not extend much beyond 4° S. In part this is due to lack of suitable physiographic conditions, but it may also be related to ocean currents and dispersal as well as water temperatures. This is an issue that needs further clarification. In Africa mangal occurs south to 32° on the east coast but on the west coast the limit is 10° S, again perhaps related to water temperatures. In Australia and New Zealand *Avicennia* occurs south to 37°.

---

[6] Mangrove communities.

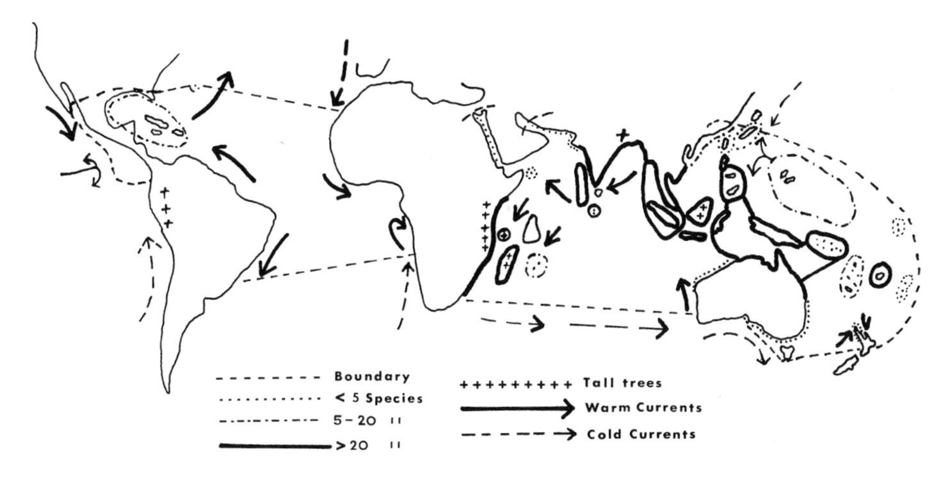

Fig. 2. Distribution of Mangrove (After CHAPMAN)

The mangrove vegetation of the world falls naturally into two great groups,—the Old World mangal and the New World and West African mangal. The distribution of these two groups is illustrated in Fig. 2. The Old World mangal contains a large number of species, about 60, the number varying according to the interpretation of what constitutes mangrove. In contrast there are only about 10 species in the New World group. It can, therefore, be argued that the original centre of distribution must have been in the Old World (CHAPMAN, 1944, 1970), probably in the Indo-Malesian area where the greatest number of species are to be found.

In Africa there is a sharp boundary between species of the two groups on the east and west coasts. In the Pacific only one species of the New World, *Rhizophora mangle*, has extended westwards to Fiji where it grows along with Old World species. Their absence from intervening islands has led to the suggestion that the seedlings were carried to Fiji by primitive man in his early wanderings, because the use of the bark for tanning rope and sails extends back into antiquity (CHAPMAN, 1970).

Just as the salt marsh vegetation can be subdivided into areal groups based mainly upon floristic differences, so the mangal vegetation can be subdivided into groups. These are as follows:

I. New World Group

This can perhaps be subdivided essentially on physiographic differences. The Florida and Gulf coast mangrove represent a region where salt marsh, fresh water swamp and mangrove intermingle on a subsiding coast line, and where fires and hurricanes appear to be significant ecological factors. Lagoon mangrove of Mexico, Venezuela, Central America and Central West Africa is another subdivision characterized by seasonal mixing of salt and fresh water. The other two subdivisions are coastal mangrove at the mouths of rivers or in protected bays and coral cay mangrove. The former exhibit some floristic differences as between the east and west coasts of Central America. *Pelliciera rhizophorae*, for example, is restricted to the Pacific. There are also some floristic and ecological distinctions as between West Africa and Atlantic America. West African mangal is essentially restricted to lagoons, bays and estuaries and in all these a greater part is played by *Rhizophora racemosa* and *R. harrisonii* as compared with *R. mangle*, and *Avicennia germinans* is replaced by *A. africana*. These differences are perhaps sufficient justification to treat the West African mangal as a separate group.

II. East African Group

The major development here is on the shores of Tanzania and Mozambique where many of the characteristic Old World species are present. The species decrease in number in the north and only two (*Avicennia marina* and *Rhizophora mucronata*) penetrate the Red Sea.

III. Indian Group

With better environmental conditions a greater number of species and additional ecological communities are present. Mangrove in the Bay of Bengal can be regarded as forming a distinct subgroup. Here monsoonal and snow melt floods from the Ganges, Bramaputra and Irrawaddy result in the swamps virtually becoming freshwater for parts of the year. These conditions favor *Heritiera minor* and the species occupies large areas which are commercially valuable.

IV. Indo-Malesian Group

These floristically are the richest and they comprise some of the most magnificent mangrove forests in the world. The richness of the flora suggests that this area represents the one in which mangroves first developed and from which they have spread to other parts of the world (Chapman, 1975). In these forests *Bruguiera cylindrica* and *B. parviflora* can form important communities.

V. Australasian Group

In Australia *Rhizophora stylosa* plays a more important role than other species of the genus and the pantropical mangrove fern *Acrostichum aureum* is replaced by *A. speciosum* (MacNae, 1968). *Avicennia marina* is represented by var. *resinifera*, and in the south and also in New Zealand this variety mingles with salt marsh communities.

VI. Philippines, New Guinea, and Oceania Group

These swamps are characterized by the presence of two species of *Camptostemon* (Bombacaceae) as well as by a greater role played by *R. stylosa*. Among the epiphytes there are the myrmecophilous plants *Hydnophytum* and *Myrmecodia*.

Whilst species of *Avicennia* normally mark the northern or southern limits of mangrove, in Japan *Kandelia candel* takes over this role.

The major salinity problem posed to both salt marsh and mangrove swamp is tidal inundation. On the salt marsh most, if not the entire, plant can be submerged at spring tides. In mangrove it is only the roots, pneumatophores and trunk base that become inundated, except in the case of seedlings when the entire plant can be submerged.

F. Distribution of Inland Salt Marshes and Salt Deserts

Inland salt marshes and salt deserts represent a distinct type of habitat where the salinity problem is essentially related to ground waters and also where salts of sodium other than sodium chloride are involved, as well as salts of magnesium. The dominant vegetation in these habitats is essentially shrubby. These inland marshes and salt deserts can be divided into regional groups in the same way as the maritime salt marshes and mangrove swamps. The division is based essentially upon floristic differences, but there is no one centre of distribution as in mangrove vegetation, nor does the method of salt desert formation appear to be so important as does the method of salt marsh formation. The main groups are as follows (CHAPMAN, 1972, 1974):

I. Inland European Group

Whilst there is general similarity throughout, they can be subdivided biogeographically into northern and south-eastern subgroups. The northern subgroup is clearly related to the northern European maritime marshes because some species are common. The grass *Puccinellia distans* is a characteristics species of this subgroup. The other subgroup is representend by an influx of eastern species such as *Crypsis aculeata*, *Aster tripolium* var. *pannonicus*, *Lepidium cartilagineum* etc.

II. Inland Asian Group

These are not so well known as some of the others, and when more information becomes available the present three subgroups may need to be multiplied and one or more may even require elevation to group status. The Aralo-Caspian subgroup exhibits some affinities with the south east European subgroup, but there are species such as *Kalidium foliatum*, *K. caspicum*, *Halostachys caspia* and *Halocnemum crassifolia* that are typical of these salt deserts. The Iraq-Central Asian subgroup has some affinities with the North African subgroup (see below), but it does possess some characteristic species, e.g. *Suaeda vermiculata*, *S. palestina*, *Seidlitzia rosmarinus* (ZOHARY, 1963; HABIB et al., 1971). The third subgroup — East Asian — is least known, but seems to be typified by species of *Atriplex* and *Artemisia*.

III. African Group

This is conveniently divided into northern, eastern and southwestern subgroups, though as further information becomes available additional subdivisions may be desirable. The north African subgroup contains western elements in the Moroccan area, but as one travels eastwards there is an increasing infiltration of eastern species. The most widespread community is dominated by *Salicornia fruticosa* and *Sphenopus divaricatus*. The East African subgroup shares some species in common with the north African subgroup but there are other species, e.g. *Halopeplis perfoliata*, *Aeluropus lagopoides* that are typical of the area. The S. West African subgroup is least known but around saline lakes *Scirpus robustus* and *S. spicatus* are the primary colonists.

IV. Inland North American Group

Some species that occur are found also on maritime salt marshes of either the Pacific or Atlantic coasts, but there are others, *Salicornia utahensis*, *Allenrolfea occidentalis*, *Sarcobatis vermiculata*, *Atriplex confertifolia*, that are characteristic of the group (HUNT and DURRELL, 1966; DODD and COUPLAND, 1966; CUSICK, 1970; UNGAR, 1970; BRADLEY, 1970). Based upon published accounts it would seem that the group can be subdivided into a north-western subgroup, characterized by *Salicornia rubra* and *Puccinellia nuttallii*, and a south western subgroup characterized by *Sporobolus airoides* and *Tamarix pentandra*.

V. South American Group

Characteristic species of inland saline areas are *Heterostachys ritteriana*, *Allenrolfea patagonica*, *Spartina montevidensis* and *Salicornia gaudichaudiana*. At present insufficient data does not permit one to determine whether the group should be subdivided.

VI. Australian Inland Group

These are mainly characterized by species of *Atriplex* and *Arthrocnemum halocnemoides*, the latter also occurring on the maritime salt marshes (p. 16).

G. Economic Uses

Saline areas must have been familiar to man for a very long time. Because so much glycophytic land has been available there has been no urge to utilize land that clearly was not favorable for plant growth. At some period in tropical areas local inhabitants must have found that the landward parts of mangrove swamps could be used for the growing of crops whilst the timber could be used for house construction, fish net poles and as a fuel. Similarly, at an early stage, the use of

Rhizophora bark as a source of tannin for use with rope and sails must have emerged.

At present mangrove soils in tropical Africa are used for growing sugar, rice, bananas and rubber (GREWE, 1941), but although mangrove soils in India have been examined for rice culture (SIDHU, 1963) they have not so far been used extensively. Where the soil is more sandy, coconuts are the preferred crop. The mangrove forests are regarded in some parts of the world as very important commercial resources and are generally under state forest service control, e.g. E. Africa, Bay of Bengal, Indo-Malesia, Philippines. The forest managerial practice varies, depending on the use to be made of the timber. In Johore, for example, where maximum output of firewood and charcoal is the objective, management aims to maintain the highest possible production of *Rhizophora* in the stands and the working plan has been changed from a 40 to a 20 year cycle (EDINGTON, 1963).

The principal sources of tannin are the mangrove swamps of E. Africa, India, Indonesia and the Philippines with lesser quantities from other areas. Although the Chinese and Arabs had used mangrove bark for tanning for a very long time, tropical America provided the first source of material for the European market. Despite this it appears that little effort has been made to exploit the new world mangal systematically nor is there any attempt at artificial regeneration (WEST, 1963). Production of tannin on the East African coast commenced in 1890 and it has been estimated that the potential annual harvest could amount to 10000 tons dry weight. The Andaman Islands in the Indian Ocean could yield 20000 tons annually (SAHNI, 1958). In Indonesia not less than 10000 tons of bark are required annually for tannin production. Malaya is the principal source of mangrove charcoal and the amount produced must be very considerable. In 1948 over 28000 tons were made in Matang alone. In West Africa and the Bay of Bengal mangrove is one of the main sources of firewood, whilst in E. Africa mangrove wood is annually exported for building purposes to Somaliland, Arabia and the Persian Gulf. In the Bay of Bengal, Sundri *(Heritiera minor)* is the principal timber tree with a very wide variety of uses, whilst *Excoecaria* is being increasingly employed for pulp production.

The extensive uses to which mangrove timber has been put represents what is probably the optimum use of saline soils for commercial purposes.

The principal use of maritime salt marshes has been for grazing of domestic animals and then later as a source of turf for bowling greens. The grazing usage has existed for a long period on the sandy north sea and Atlantic marshes of Britain and Europe. The Swedish marshes in particular have a long history of utilization (DAHLBECK, 1945). For over 500 years they have been grazed by goats, sheep and cows. Up to the eighteenth century areas of turf were also removed for thatching of cottages. In New England and Nova Scotia the salt marshes were regularly mown each year for hay and the staddles on which the ricks were built were certainly evident some years ago. In Europe annual species of *Salicornia*, e.g. *S. stricta*, were regularly collected as a vegetable and sold in markets, and at least one firm is still interested in the use of the species as a pickle. The marshes have always been a haven for duck and wildfowl and this is at least one reason for protests when marshes are converted to other purposes.

Marshlands, because of their steady rise in height, have in many countries long been regarded as potential assets. The marsh generally reaches a height where the adjacent land owner considers it an economic proposition to construct a sea wall around it, drain the area, desalinize it, usually by natural rainfall, and turn it into pasture. Much of the fenlands of England have clearly been salt marsh at one time and successive lines of sea walls record historically the reclamation of salt marsh in England from Roman times onward. Successful reclamation involves retaining the sea walls in good condition and repairing immediately any breach that may be made.

In more recent years demands by man have resulted in the reclamation of inland saline soils. The only natural use of such soils is adjacent to salt lakes where impoundments provide natural salt deposits by evaporation. Inland saline soils are generally reclaimed by the addition of gypsum, though on Hungarian soils calcium nitrate has been found equally satisfactory (SZABOLCS, 1965). Under certain conditions sulphur and soil conditioners are valuable aids. Deep plowing has also proved successful in Canada (CAIRNS, 1971), Hungary and Russia and can be followed by the sowing of grass and clover. Agriculturists can choose between gypsum or deep-plowing as a technique and it is likely that the economic factor will determine the choice. Once reclamation has been undertaken close attention must be given to the first crops to be planted. Provided they are selected carefully good growth will ensue, and because these first crops remove still more salt from the soil, eventually virtually any crop can be grown.

In Australia some of the natural saline areas have been restored by treatment with gypsum and contour plowing. Initially salt-tolerant native species, *Kochia* spp., *Atriplex* spp. are sown and then later buffalo grass *(Stenotaphrum)*, (STONEMAN, 1958). In those areas of Queensland where secondary salinization has taken place the planting of rice and irrigated mixed pastures has helped to restore the chemical balance. *Phalaris tuberosa*, woomera rye grass and subterranean clover are sown and irrigated for 3–5 years and then rice crops are planted for the next two years. Tile or open drains are also used to assist in salt leaching.

In India, soils that are moderately saline can be successfully and economically reclaimed by prolonged leaching followed by crops of rice. More heavily salinized soils in both India and Pakistan can be economically reclaimed if cheap irrigation water is available for leaching out the salt. They are then often planted with a crop of *Sesbania aculeata* or *S. aegyptiaca*. In the Punjab, success has been achieved by scraping off the highly saline surface layers and planting with *Cynodon dactylon*. The most heavily salinized soils are generally uneconomic to reclaim though some success has been achieved by planting salt tolerant tree species (*Prosopis julifera*, *Acacia arabica* etc.) in holes after the hard pan has been broken and manure added (RAHEJA, 1966).

In Iran and Iraq the saline soils generally contain sufficient gypsum so that only extensive irrigation, flooding and leaching is required for successful reclamation. In Egypt moderately salty land is reclaimed by prior leaching and the application of gypsum for a period of at least three years. Such soils are normally planted with rice (RAHEJA, 1966). No attempt has been made to reclaim the heavily salinized soils.

Where man has been responsible for secondary salinization it means that further attention must be given to water management practices to make them more efficient. In some cases the economics of the situation may need study, and it could prove more profitable to let the areas revert to their original natural state. Where reclamation of natural saline soils is practiced, care must be taken to see that any monoculture introduced is not likely to make the final situation worse. Finally, reclamation of maritime soils needs careful consideration because of their natural high productivity, which should, perhaps, not be disturbed. Reclamation may not, in fact, be economically desirable.

References

AGAEV, B. M.: Preliminary results of the reclamation of soda-saline soils in the Karabah Plain, Azerbaidzhan. In: Symposium on sodic soils, Budapest, Hungary, 1964. Agrokem. Talajtan **14** (Suppl.), 189–194 (1965).

ANON: Man in the living environment. Inst. of Ecology Rept. of the Workshop on Global Ecological Problems. Inst. Ecol. Pub. (1972).

ASGHAR, A. G.: Report on irrigation practices of Pakistan. Third Region. Irrig. Practices leadership seminar, pp. 55–64 (1960).

BARBOUR, M. G.: Is any Angiosperm an obligate halophyte? Am. Mid. Nat. **84** (1), 105–120 (1970).

BOUMANS, J. H., HUSBOS, W. C.: The alkali aspects of the reclamation of saline soils in Iraq. Neth. J. Agric. Sci. **8** (3), 225–235 (1960).

BOYKO, H.: Introduction 1–22. Salinity and aridity: New approaches to old problems. The Hague: Junk 1966.

BRADLEY, W. G.: The vegetation of Saratoga Springs National Monument, California. Southwest Nat. **15** (1), 111–129 (1970).

CAIRNS, R. R.: Canadian solonetz soils and their reclamation. Agrokem. Talajtan **18**, 233–37 (1969).

CAIRNS, R. R.: Effect of deep plowing on the fertility status of black solonetz soils. Can. J. Soil Sci. **51**, 411–14 (1971).

CHAPMAN, V. J.: Cambridge university expedition to Jamaica. J. Linn. Soc. Lond. (Bot.) **52**, 407–533 (1944).

CHAPMAN, V. J.: Vegetation and salinity, 23–42, In: BOYKO, H. (Ed.): Salinity and aridity: new approaches to old problems. The Hague: Junk 1966.

CHAPMAN, V. J.: Mangrove phytosociology. Trop. Ecol. **11** (1), 1–19 (1970).

CHAPMAN, V. J.: Salt marshes and salt deserts of the world. Salt Marsh Symp. AIBS meeting (1972). In: Ecology of Halophytes: Ed. Reimold and Queen, Acad. Press 1974.

CHAPMAN, V. J.: Salt Marshes and salt deserts of the world, 2nd ed. Lehre: Cramer Verlag 1974.

CHAPMAN, V. J.: Mangrove vegetation. Lehre: Cramer Verlag 1975.

CUSICK, A. W.: An assemblage of halophytes in northern Ohio. Rhodora **72**, 285–86 (1970).

DAHLBECK, N.: Strandwiesen am südöstlichen Oresund. Act. Phytogeogr. Suec. **18** (1945).

DODD, J. P., COUPLAND, R. J.: Vegetation of saline areas in Saskatchewan. Ecol. **47**, 958–967 (1966).

EDINGTON, P. W. J.: Working plan for the South Johore Mangrove working circle 2nd Rev. 1960–64 Johore (1963).

GILLNER, V.: Vegetations- und Standorte-Untersuchungen in den Strandwiesen der Schwedischen Westküste. Act. Phyt. Suec. **43**, 1–198 (1960).

GREWE, F.: Afrikanische Mangrovelandschaften, Verbreitung und wirtschaftsgeographische Bedeutung. Wissensch. Veröffentl. Deut. Mus. Landwerk. N. F. **9**, 105–177 (1941).

HABIB, Z. M., AL-ANI, T. A., AL-MUFTI, M. M., AL-TAWIL, B. H., TAKESSIAN, B. A.: Plant indicators in Irak. I. Native vegetation as indicators of soil salinity and water-logging. Plant Soil **34** (2), 405–415 (1971).

HAYWARD, N. E.: Plant growth under saline conditions. Reviews of research on problems of utilization of saline water, pp. 37–72. Paris: UNESCO 1954.
HUNT, C. B., DURRELL, L. W.: Plant ecology of Death Valley. Geol. Surv. Prof. Pap. **509**, 1–68 (1966).
KELIER, P., HENIS, Y.: The effect of sodium chloride on some physiological groups of microorganisms inhabiting a highly saline soil. Israel J. Agr. Res. **20** (2) 71–75 (1970).
LÜKEN, H.: Saline soils under dry-land agriculture in southeastern Saskatchewan (Canada) and possibilities for their improvement. I–II. Plant and Soil **17** (1), 1–67 (1962).
MACNAE, W.: A general account of the fauna and flora of mangrove swamps and forests in the Indo-West-Pacific region. Advan. Mar. Biol. **6**, 73–269 (1968).
MURATOVA, V. S.: Solonetzes of the Milsk alluvial plain (Kura-Araks Lowland). Pochvovedenie **1959** (9), 1041–53 (1961).
OPREA, C. V.: The genesis, development and amelioration of Alkali soils on the western lowlands of the Rumanian People's Republic. In: Symposium on Sodic Soils, Budapest, Hungary, 1964. Agrokem. Talajtan **14** (Suppl.), 183–188 (1965).
ORLOVSKIJ, N. V.: The genesis and utilization of salt-affected soils in Siberia. In: Symposium on Sodic Soils, Budapest, Hungary, 1964. Agrokem. Talajtan **14** (Suppl.), 155–174 (1965).
OVDENICHAROV, I. N.: Solonetzic soils of the Frakia-depression in Bulgaria. Pochvovodenie **1959** (9), 1090–99 (1961).
POLLACK, G., WAISEL, Y.: Salt secretion in *Aeluropus litoralis* (Willd.) Parl. Ann. Bot. (Lond.) N.S. **34** (137), 879–888 (1970).
RAHEJA, P. C.: Aridity and salinity (A survey of soils and land use), 43–127. In: BOYKO, H. (Ed.): Salinity and aridity: new approaches to old problems. The Hague: Junk 1966.
SAHNI, K. C.: Mangrove forests of Andaman and Nicobar Islands, In: Symposium in Mangrove vegetation. Sci. and Cult. **23**, 330 (1958).
SCHOLANDER, P. F.: How mangroves desalinate water. Physiol. Plant. **21**, 251–261 (1968).
SIDHU, S. S.: Studies on the mangroves of India. I. East Godavari Region. Indian For. **89**, 337–351 (1963).
STONEMAN, T. C.: Salt land program for Autumn. West. Austr. Dept. Agric. J. Ser. **7** (3), 359–360 (1958).
STROGONOV, B. P.: Physiological basis of salt tolerance of plants. Jerusalem: Israel Prog. Sci. Transl. 1964.
SZABOLCS, J.: Salt affected soils in Hungary. Agrokem. Talajtan **14**, (Suppl.), 275–290 (1965).
TEAL, J. M., KANWISHER, J. W.: Gas transport in the marsh grass, *Spartina alterniflora* (Gramineae). J. Exp. Bot. **17** (51), 355–361 (1966).
TEAL, J. M., KANWISHER, J. W.: Total energy balance in salt marsh grasses. Ecol. **51** (4), 690–695 (1970).
TEAL, J. M., TEAL, M.: Life and death of a salt marsh. Boston: Atlantic, Little, Brown and Co. 1969.
THOM, B. G.: Mangrove Ecology and Deltaic geomorphology, Tabasco, Mexico. J. Ecol. **55**, 301–343 (1967).
UNGAR, I. A.: Species-soil relations on sulphate dominated soils of South Dakota. Am. Mid. Nat. **83**, 343–357 (1970).
WEIHE, K. VON: Beiträge zur Ökologie der mittel- und westeuropäischen Salzwiesenvegetation (Gezeitenküsten). 1. Methodik, Standorte und vergleichende morphologische Analyse. Beitr. Biol. Pflanzen **39** (2), 189–257 (1963).
WEST, R. C.: Mangrove swamps of the Pacific Coast of Colombia. Ann. Ass. Am. Geog. **46**, 98–121 (1963).
ZOHARY, M.: On the geobotanical structure of Iran. Bull. Res. Counc. Isr., Sect. D. (Suppl.) **11**, 1–112 (1963).

CHAPTER II

Problems of Salinity in Agriculture

D. L. CARTER

A. Introduction

Millions of hectares of land throughout the world are too saline to produce economic crop yields, and more land becomes nonproductive each year because of salt accumulation. Salinity problems in agriculture are usually confined to arid and semiarid regions where rainfall is not sufficient to transport salts from the plant root zone. Such areas comprise 25% of the earth's surface (THORNE and PETERSON, 1954). Salinity is a hazard on about half of the irrigated area of the western USA (WADLEIGH, 1968) and crop production is limited by salinity on about 25% of this land (WADLEIGH, 1968; THORNE and PETERSON, 1954; BOWER and FIREMAN, 1957). The occurrence of salinity is similar in the arid regions of western Canada, the high plains of Mexico and the Pacific slopes of South America. Salt affected soils are also extensive in South Africa, Rhodesia, Egypt, Marocco, and Tunisia. Only small areas of salt affected soils occur in Europe, but extensive areas are present in Asia (THORNE and PETERSON, 1954). In general, it can be concluded that salinity problems are found in all countries having areas where arid or semiarid climates exist.

Soil salinity problems are present in nearly every irrigated area of the world and also occur on non-irrigated croplands and rangelands. CARTER et al. (1964) reported that approximately 25% of the non-irrigated land in the Lower Rio Grande Valley of Texas is highly saline. The saline soils are interspersed among non-saline soils so that farmers must plant and cultivate saline areas along with non-saline areas. Thus, farmers may actually harvest only about 75% of the land area they farm. Similar problems face farmers throughout the world, adding to operational costs and making farming less practical.

The actual total area of salt affected soils throughout the world is not known, but it is large. A recent survey indicated that the irrigated areas of 103 countries totaled 203 million hectares (ANONYMOUS, 1970). If 25% of this land is saline, then there are 50 million hectares of irrigated, salt affected soils. In addition, there are large areas of non-irrigated salt affected soils. Therefore, the salinity problem in agriculture is extensive and important.

The importance of the salinity problem varies among countries. AYERS et al. (1960) reported that salt affected soils of Spain constitute only a small percentage of the total arable lands, but they estimated that 250000 hectares are affected to the extent that crop yields are depressed. In contrast, the 17 Western States in the USA have a total of approximately 8 million acres of salt affected soils (BOWER and FIREMAN, 1957). It is not the purpose of this chapter to discuss salt affected soils of each nation, but rather to discuss the nature of the salinity problems in agriculture, and to some extent, what can be done about these problems. A useful bibliography of papers published on salt affected soils and their management is available (CARTER, 1966).

Salt affected lands can usually be made productive by reclamation and better management if resources are available, but reclamation cost greatly increases production costs. In many cases, projected reclamation costs far exceed expected returns from the land, and reclamation is not practical. In other cases resources, such as water for leaching and good quality water for irrigation, are not available, and reclamation cannot be accomplished.

B. Causes of Salinity

Water evaporates in a pure state, leaving salts and other substances behind. As water is removed from the soil by evapotranspiration (ET), the salt concentration in the remaining soil solution may become 4–10 times that in the irrigation water within 3–7 days after irrigation (PETERSON et al., 1970). Each irrigation adds some salt to the soil. How much is added depends upon the amount of water entering the soil and the salt concentration in the water. This salt remains in the soil and accumulates unless it is leached away by water applied in excess of crop requirements. The trend of salt accumulation in soil irrigated with waters having four different salt concentrations is illustrated in Fig. 1. It is assumed that there is no leaching and that each irrigation replaces exactly the water removed from the soil by ET. As illustrated by Fig. 1, numerous irrigations with water of low salt content, can be applied before the tolerance limit for a crop is reached, but the crop tolerance level is reached rapidly when irrigating with water of high salt content. Sufficient leaching to remove the salt brought into the soil by each irrigation will

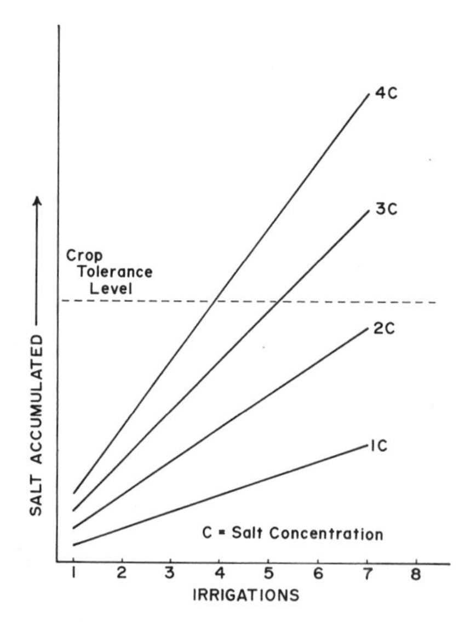

Fig. 1. Hypothetical salt accumulation in soil as related to the salt concentration in irrigation water and number of irrigations, in the absence of leaching

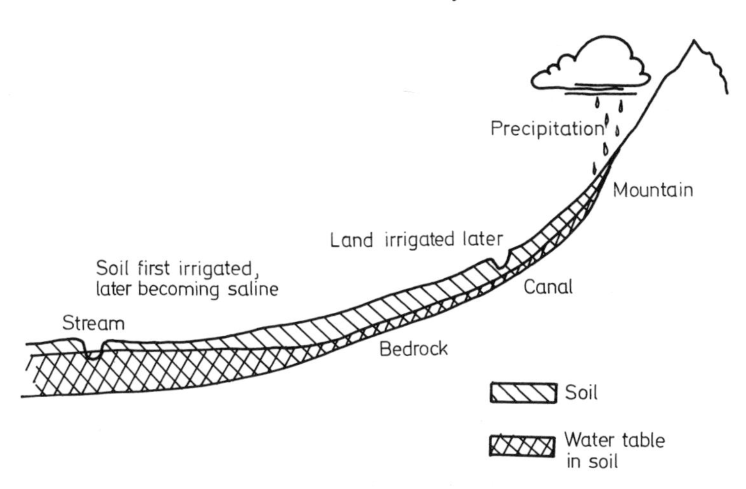

Fig. 2. Formation of salinity and high water table problems in fertile lowlands as a result of excessive irrigation at higher elevations

keep the amount of salt in the soil at the same level as it was after the first irrigation. Less leaching would not prevent salt accumulation, but would increase the number of irrigations that could be applied before the salt tolerance limit for the crop is reached. Figure 1 is a generalized illustration. It should be realized that the tolerance level varies for different crops.

Drainage water that has passed through the soil has a higher salt concentration than the irrigation water (WILCOX and RESCH, 1963; CARTER et al., 1971). Most of this drainage water returns to the natural stream or river channel, downstream from the point where the irrigation water is diverted. As a result, the salt concentration in rivers and streams in arid and semiarid regions generally increases from the headwaters to the mouth. This, in itself, creates a salinity problem for agriculture, because the salt concentration in the stream can become so high that the water cannot be used for irrigation.

Irrigating one area may cause salt problems in another. Salts may be transported from one cropped area with adequate drainage to another with inadequate drainage where they accumulate. In many irrigated valleys, this problem has seriously limited productivity on the best agricultural soils. Usually, lands adjacent to the river or stream were the first to be irrigated. Later, water distribution systems were developed to irrigate lands at a higher elevation. If too much water was applied on these higher elevation lands, drainage water from these lands caused high water tables and salt problems on those lower lands first irrigated. In many valleys, this process has caused the lands next to the river or stream to become saline marshes, useful only for wildlife and waterfowl habitat. Examples of these processes are evident in most irrigated valleys of the western U.S. Natural precipitation has caused the same problem in some areas. The situation is illustrated in Fig. 2. In some cases, artificial drainage in the low land and better water management on the entire area can return the soils adjacent to the stream to good productivity.

Although salt accumulation resulting from ET is the primary cause of salinity problems in agriculture, there are other sources of salt. Some soils naturally contain sufficient salt to limit or prohibit production of economic crops. Some of these soils were derived from saline parent materials, and some contain natural salt deposits. Old lakebed soils, for example, often contain salt deposits. Some soils have received sufficient salt from sea spray to become saline or contain naturally high salt levels (YAALON, 1963). Another salt source is the soil. As water from precipitation or irrigation passes through the soil, natural salt-bearing minerals are very slowly dissolved. As the dissolved salts are transported away by water, the equilibrium between the solid and dissolved phases shifts and more salts are dissolved. Salts from this source are transported to an area where the water is evaporating and the salts are concentrated. Although salts dissolve slowly from most soil minerals, 2.2 metric tons or more per hectare can be dissolved in a year from some calcareous silt loam soils (CARTER et al., 1971).

Ancient natural salt deposits in the soil, usually referred to as fossil salts, are found throughout arid and semiarid regions. Saline materials also underlie some soils such as the Mancos Shale in Colorado (SKOGERBOE and WALKER, 1972). Water that comes in contact with these salts becomes highly saline. If this saline water reaches a surface stream or a groundwater acquifer being used for irrigation, severe salt accumulations can result that seriously limit crop production. There are also numerous natural saline groundwater acquifers. When water is pumped from these for irrigation, salts are brought to the surface where they accumulate and damage crops. As the demand for water increases, the tendency to develop groundwater sources also increases. This results in using more medium and high-salt groundwaters, thus intensifying the salinity problem.

Phreatophytes growing along canals and drains create an especially serious problem in water-short areas. These plants use water that may be needed for irrigation, leaving the salt behind so that the salt concentration in the remaining water is increased and the quality of the water is impaired for irrigation. Phreatophytes are found along most open, unlined canals and drains of irrigated areas. Seepage from canals also frequently causes high water tables, increases soil salinization, and encourages growth of the non-economic plants. Lining canals or installing pipelines can reduce these kinds of problems, but such practices are costly.

In summary, salinity problems in agriculture arise from many sources, both natural and man-caused. Salts are transported by water, and become problems when and where the water evaporates. Water evaporation from the soil leaves salts at or near the soil surface. Water evaporation through plants leaves salts near the point of water absorption by plant roots.

C. Coping with Salinity Problems

I. Crop Selection

Plants species differ in their tolerance to total salts and to specific ions. Also, crops that may be highly tolerant at one growth stage may be sensitive during another stage. Generally, plants are most sensitive to salinity during germination or early

Table 1. The EC_e at which 10, 25, and 50% yield reductions can be expected for various agricultural crops. (Adapted from BERNSTEIN, 1964)

	Percent yield reduction		
	10	25	50
Field Crops			
Barley	11.9	15.8	17.5
Sugarbeets	10.0	13.0	16.0
Cotton	9.9	11.9	16.0
Safflower	7.0	11.0	14.0
Wheat	7.1	10.0	14.0
Sorghum	5.9	9.0	11.9
Soybean	5.2	6.9	9.0
Sesbania	3.8	5.7	9.0
Rice	5.1	5.9	8.0
Corn	5.1	5.9	7.0
Broadbean	3.1	4.2	6.2
Flax	2.9	4.2	6.2
Beans	1.1	2.1	3.0
Vegetable Crops			
Beets	8.0	9.7	11.7
Spinach	5.7	6.9	8.0
Tomato	4.0	6.6	8.0
Broccoli	4.0	5.9	8.0
Cabbage	2.5	4.0	7.0
Potato	2.5	4.0	6.0
Corn	2.5	4.0	6.0
Sweetpotato	2.5	3.7	6.0
Lettuce	2.0	3.0	4.8
Bellpepper	2.0	3.0	4.8
Onion	2.0	3.4	4.0
Carrot	1.3	2.5	4.2
Beans	1.3	2.0	3.2
Forage Crops			
Bermudagrass	13.0	15.9	18.1
Tall wheatgrass	10.9	15.1	18.1
Crested wheatgrass	5.9	11.0	18.1
Tall fescue	6.8	10.4	14.7
Barley hay	8.2	11.0	13.5
Perennial rye	7.9	10.0	13.0
Hardinggrass	7.9	10.0	13.0
Birdsfoot trefoil	5.9	8.1	10.0
Beardless wildrye	3.9	7.0	10.8
Alfalfa	3.0	4.9	8.2
Orchardgrass	2.7	4.6	8.1
Meadow foxtail	2.1	5.5	6.4
Clovers, alsike and red	2.1	2.5	4.2

seedling growth. Some crops, such as rice, are also sensitive during flowering and seed set (PEARSON et al., 1966). Well-established plants are usually more tolerant than new transplants (BERNSTEIN, 1964). Therefore, crop selection is an important management decision in salinity affected areas. Severe salinity problems may preclude growing salt sensitive crops and may require special practices to produce economic yields of salt tolerant crops. Mild salinity problems may require special practices only for crop establishment.

The United States Salinity Laboratory Staff has conducted extensive research on the salt tolerance of agricultural plant species, and combined their results with those of other researchers throughout the world to compile salt tolerance listings for different plant species (U.S. Salinity Laboratory Staff, 1954; BERNSTEIN, 1964). In these listings the tolerable level of salinity is based on the electrical conductivity of saturated soil extracts for which the symbol is EC_e and the units are millimhos per centimeter (mmhos/cm). The EC_e is related to the number of electrical conducting ions in solution, and an EC_e of 1.0 corresponds to a total salt concentration of approximately 640 parts per million. The salt concentration in the soil solution at field capacity will be twice that in the saturated extract. A listing of salt tolerance values adapted from data reported by BERNSTEIN (1964) for field, vegetable and forage crops is presented in Table 1. These data are based on EC_e measurements made on the saturated soil extracts during the period of rapid plant growth and maturation, from late seedling stage onward. Some plants are sensitive to specific ions such as boron (WILCOX, 1960), chloride, and sodium (BERNSTEIN, 1964), but this will not be dicussed in this chapter.

II. Crop Stand Establishment

Because many plant species are most sensitive to salinity during germination or early seedling development, several cultural and management practices have been developed to enhance stand establishment. One practice is to irrigate lightly each day after seeding with a solid set sprinkler system until the stand is established and then convert to furrow or less frequent sprinkler irrigation. Another technique is to leach salts from surface soils just before planting to allow stand establishment before salts can accumulate enough to interfere with germination or to damage seedlings. This approach is not always successful because if evaporative demand is high, water moves rapidly to the soil surface, evaporates, and leaves salt behind. A third cultural practice is to bed the area to be seeded so that salts accumulate at the ridge tops, and then seed on the slope between the furrow bottom and ridge top (Fig. 3) (BERNSTEIN and FIREMAN, 1957; BERNSTEIN, 1964). This can be done with double row or single row seeding. Trickle irrigation is a new practice that appears to have some merit for establishing stands and for growing the crop to maturity (BERNSTEIN and FRANCOIS, 1973; GOLDBERG et al., 1971; GOLDBERG et al., 1971). This practice keeps a zone low in salt around the plant roots or seed.

Fortunately, in many areas, precipitation between cropping seasons is sufficient to leach salts from surface soils and to allow crop stand establishment before salts accumulate again to harmful concentrations. In such areas, evaporation between crop seasons is usually low, and the precipitation effectively leaches salts downward in the soil without much upward return to the soil surface. Leaching by natural precipitation will be most effective if the soil water content is high when the precipitation occurs, so that the water from precipitation will not all be

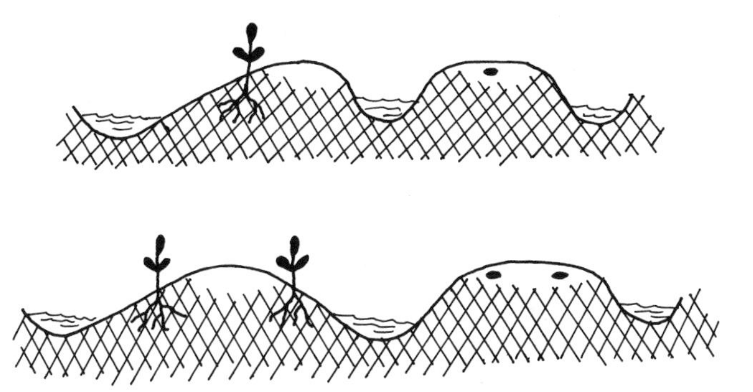

Salts are moved to the ridge tops and accumulate there as water evaporates. Seed placed in the salt accumulation zones may not germinate or seedlings may die because of salt effects. Seed placed along the slope is less likely to be affected by salt accumulations. This method can be used with both single and double row systems.

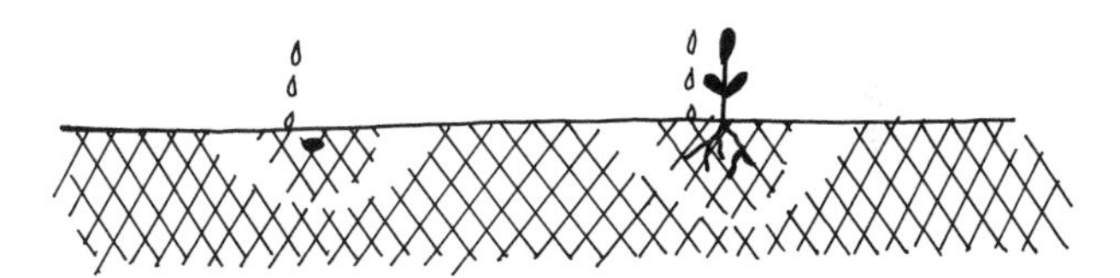

Drip irrigation maintains a low salinity zone around the seed and allows germination and plant establishment.

Salt accumulations

Fig. 3. Bedding soil so that salts accumulate at the high point. Seeding on the sloping soil between the water line in the furrow and the high point allows stand establishment. Drip irrigation moves salt away from the seed and root zone

used for filling the soil root zone, but will force water from it. Although it is not widely practiced, leaching during a cool noncropping season, when evaporation is low, is very effective if good quality water is available. In some areas the non-cropping season is during the warmest and driest time of the year. Under such conditions, the approach described above is not usually beneficial.

Several other practices have been used to enhance seedling establishment and increase production. CARTER and FANNING (1965) applied a cotton bur mulch to reduce evaporation and enhance leaching, and then seeded grain sorghum through the mulch. A satisfactory crop was produced by this method. These investigators had shown earlier that cotton bur and wood chip mulches effectively reduced evaporation, increased water intake, and increased leaching of salts (FANNING and CARTER, 1963; CARTER and FANNING, 1964). RASMUSSEN et al. (1972) have demonstrated that deep plowing of some salt affected soils brings about a complete reclamation. The success of deep plowing depends upon the texture and lime or gypsum content of the subsoil. High clay content in the subsoil limits the success of this practice.

III. Leaching Requirements

The excess water applied periodically to leach salts from the root zone is known as the leaching requirement (U.S. Salinity Laboratory Staff, 1954; WILCOX and RESCH, 1963; THORNE and PETERSON, 1954). Where soils are only slightly saline and the electrical conductivity of the irrigation water (EC_e) is below about 0.85 mmhos/cm, normal irrigation practices usually provide for sufficient leaching. What really happens is that the inefficiency of the irrigation practice provides sufficient excess water to meet the leaching requirement. Improving the irrigation efficiency increases the importance of taking into account the leaching requirement. If salt has accumulated in the root zone to near the tolerance level for the more tolerant plants, or if the irrigation water has an EC_e of 1.5 mmhos/cm or greater, leaching requirements are large and normal supplies of irrigation water may be insufficient to meet them. The specific effects of sodium on the structure and properties of the soil should also be considered when leaching.

IV. Drainage

Adequate drainage is essential for salinity control. The buildup of the groundwater table with subsequent movement of water to the soil surface to meet evaporative demands is a primary cause of soil salinity. Salinity resulting from high water tables can often be prevented or eliminated by applying less water or by proper drainage. Artificial drainage is costly, however, and drainage requirements and attendant costs should be carefully considered before a new area is brought under irrigation.

D. The Salt Balance Concept

The salt balance for any given land area or soil unit can be expressed by the following equation:

$$Sp + Si + Sr + Sd + Sf = Sdw + Sc + Sppt.$$

Where:

Sp = salt in natural precipitation falling upon the area,
Si = salt in the irrigation water diverted into the area,
Sr = residual salts in the soil,
Sd = salt dissolved into solution from soil minerals,
Sf = salt in applied fertilizers,
Sdw = salt in the drainage water from the area,
Sc = salt taken up by the crop and removed,
Sppt = salt chemically precipitated in the soil.

The salt in natural precipitation, Sp, is usually extremely small and can be ignored. In some areas, however, occasional storms do deposit sufficient salt to cause problems; for example, when strong winds have brought ocean spray inland or when dusts from salt flats have become airborne and then precipitate elsewhere. Si depends upon the salt concentration in the irrigation water, and the amount of water diverted into the area. Si is large for saline waters, and in many instances, it is the most important salt input factor. Sr, in some instances, is a large quantity such as found in arid land not previously irrigated, or where fossil salt deposits occur, but in soils used in agriculture the value of Sr is usually small. Sd varies with the kinds of minerals in the soil, and depends upon the amount of water passing through the soil and the salt concentration in that water. Sf is usually small and can generally be ignored. Sdw is usually a large output component particularly where irrigation waters are of poor quality and leaching is required. It may also be large for new lands brought under irrigation. Sc can be ignored in most irrigated areas because it is small in relation to other components. Sppt is a very important and complicated factor that is often not adequately evaluated. Under most irrigation practices, Sppt is small because excess water passing through the soil does not allow the salt concentration to become high enough for precipitation to occur. Ideally, Sppt should be maximized so that slightly soluble salts such as calcium carbonate and calcium sulfate will precipitate and remain in the soil. These salts when precipitated in the soil have little or no effect upon growing crops.

The salt concentration in the soil solution in the plant root zone is of primary importance. It increases between irrigations, as water is removed by evapotranspiration. Adding irrigation water decreases the salt concentration by dilution. The magnitude of the decrease depends upon the salt concentration in the irrigation water. If the salt concentration in the irrigation water is high–for example, above 1.5 mmhos/cm–the dilution will be slight, and more frequent irrigations will be required to maintain the salt concentration in the root zone below the level that would adversely affect plant growth and production. The application of more irrigation water adds more salt to the soil, requiring a greater leaching fraction to maintain a salt balance and avoid salt accumulation in the root zone.

Ideally, a salt balance should be maintained on all irrigated lands, once the salt concentration in the root zone has been reduced to the practical level that has the least adverse affects on plant growth and production. Maintaining a salt balance would maximize chemical precipitation of harmless salts; a minimum quantity of salt would be dissolved from soil minerals, and a minimum quantity of salt would be returned to river systems in drainage water. However, two important factors must be considered before water management practices are changed to maintain a salt balance: (1) the salinity tolerance of the crops grown governs the salt concentration permissible in the soil solution, below which the salt balance must be established for successful cropping, and (2) the salt concentration in drainage water will be likely to increase as irrigation practices are changed to effect a salt balance. The use of these drainage waters may be important, and increasing their salt concentration may render them unsatisfactory for that use. A third factor to consider in some areas is the disposal of animal, food processing and industrial waste effluents on the land. Some of these effluents contain large quantities of salt that will certainly have a significant impact upon the salt balance of an area.

E. Irrigation Return Flows

The increasing salt concentration in surface and groundwaters from irrigation return flows and drainage waters, is one of the most important salinity problems in agriculture. If one-half of the water applied for irrigation is used in evapotranspiration, the salt concentration in the drainage water will be twice that in the irrigation water, assuming that no salts are dissolved from or precipitated in the soil. Increasing the quantity of irrigation water applied would be likely to decrease the salt concentration in the return flows by dilution, but would probably increase the total salt loading from them by increasing Sd in the salt balance equation. Decreasing the amount of irrigation water may increase the salt concentration in the drainage waters, but it may decrease the salt loading by decreasing Sd and increasing Sppt. More research is needed to determine the influence of the amount of water applied on the Sd and Sppt factors in the salt balance equation.

Irrigating additional lands may increase the salt load in surface streams. This depends upon the type of irrigation used. Sprinkler, mist and trickle irrigation may not cause much return flow, but the salt left behind by the water used in evapotranspiration must go somewhere. Sometimes it is stored in the soil below the root zone. In some soils, tremendous quantities of salts can be stored between the bottom of the root zone and the groundwater for many years. A very efficient irrigation practice that will move water and salt downward primarily by unsaturated flow is required to accomplish this. However, the storage capacity may ultimately be filled, and then large quantities of salt can be expected in effluents for many years causing salinity problems to the future generations.

F. Conclusions

Excessive salt accumulations prevent or limit the production of economic crops on millions of hectares. These are primarily irrigated lands in arid and semiarid regions. Reducing the salt accumulations requires special management practices, which increase production costs. Excessive salt accumulations result primarily from water being removed in evapotranspiration, leaving salts behind to concentrate in the remaining water. Leaching to remove these accumulations impairs the quality of surface and groundwater in these areas, by increasing the salt concentration of the drainage waters. Each year more is understood about managing salt affected soils, but the problems are far from being solved and the need continues for new knowledge that may lead to more effective solutions of the salinity problems in agriculture.

References

ANONYMOUS: Irrigation statistics of the world. ICID Bul., International Commission on Irrigation and Drainage **48**, Nyaya Marg, Chanakyapuri, New Delhi-21, India, pp. 76–78, Jan. (1970).

AYERS, A. D., VAZQUEZ, A., DE LA RUBIA, J., BLASCO, F., SOMPLON, S.: Saline and sodic soils of Spain. Soil Sci. **90**, 133–138 (1960).

BERNSTEIN, L.: Salt tolerance of plants. USDA Agr. Inf. Bul. **283** (1964).

BERNSTEIN, L., FIREMAN, M.: Laboratory studies on salt distribution in furrow-irrigated soil with special reference to the pre-emergence period. Soil Sci. **83**, 249–263 (1957).

BERNSTEIN, L., FRANCOIS, L. E.: Comparisons of drip, furrow, and sprinkler irrigation. Soil Sci. **115**, 73–85 (1973).

BOWER, C. A., FIREMAN, M.: Saline and alkali soils. In: USDA yearbook of agriculture, soil, pp. 282–290 (1957).

CARTER, D. L.: A bibliography of publications in the field of saline and sodic soils (through 1964). USDA, ARS 41–80 (1966).

CARTER, D. L., BONDURANT, J. A., ROBBINS, C. W.: Water-soluble NO_3-nitrogen, PO_4-phosphorus, and total salt balances on a large irrigation tract. Soil Sci. Soc. Am. Proc. **35**, 331–335 (1971).

CARTER, D. L., FANNING, C. D.: Combining surface mulches and periodic water applications for reclaiming saline soils. Soil Sci. Soc. Am. Proc. **28**, 564–567 (1964).

CARTER, D. L., FANNING, C. D.: Cultural practices for grain production through a cotton bur mulch. J. Soil Water Cons. **20**, 61–63 (1965).

CARTER, D. L., WIEGAND, C. L., ALLEN, R. R.: The salinity of non-irrigated soils in the Lower Rio Grande Valley of Texas. USDA, ARS 41–98 (1964).

FANNING, C. D., CARTER, D. L.: The effectiveness of a cotton bur mulch and a ridge-furrow system in reclaiming saline soils by rainfall. Soil Sci. Soc. Am. Proc. **27**, 703–706 (1963).

GOLDBERG, D., GORNAT, B., BAR, Y.: The distribution of roots, water, and minerals as a result of trickle irrigation. J. Am. Soc. Hort. Sci. **96**, 645–648 (1971).

GOLDBERG, S. D., RINOT, M., KARN, N.: Effects of trickle irrigation intervals on distribution and utilization of soil moisture in a vineyard. Soil Sci. Soc. Am. Proc. **35**, 127–130 (1971).

PEARSON, G. A., AYERS, A. D., EBERHARD, D. L.: Relative salt tolerance of rice during germination and early seedling development. Soil Sci. **102**, 151–156 (1966).

PETERSON, H. B., BISHOP, A. A., LAW, J. P., JR.: Problems of pollution of irrigation waters in arid regions. In: LAW, J. P., WITHEROW, J. L. (Eds.): Water quality management problems in arid regions, pp. 17–27. U.S. Environmental Protection Agency, Water Pollution Control Research Series 1970.

RASMUSSEN, W. W., MOORE, D. P., ALBAN, L. A.: Improvement of a Solonetzic (slick spot) soils by deep plowing, subsoiling, and amendments. Soil Sci. Soc. Am. Proc. **36**, 137–142 (1972).

SKOGERBOE, G. V., WALKER, W. R.: Salinity control measures in the Grand Valley. In: Managing irrigated agriculture to improve water quality, pp. 123–136. Proceedings of National Conference on Managing Irrigated Agriculture to Improve Water Quality, U.S. Environmental Protection Agency and Colorado State University 1972.

THORNE, D. W., PETERSON, H. B.: Irrigated soils—their fertility and management. New York: The Blakiston Co. Inc. 392 (1954).

U.S. Salinity Laboratory Staff: Diagnosis and improvement of saline and alkali soils. USDA Hdbk. **60**, 160 (1954).

WADLEIGH, C. H.: Wastes in relation to agriculture and forestry. USDA Misc. Pub. **1065**, 112 (1968).

WILCOX, L. V.: Boron injury to plants. USDA Inf. Bul. **211** (1960).

WILCOX, L. V., RESCH, W. F.: Salt balance and leaching requirements in irrigated lands. USDA Tech. Bul. **1290** (1963).

YAALON, D. J.: On the origin and accumulation of salts in groundwater and in the soils of Israel. Bull. Res. Counc. Israel **11 G**, 105–131 (1963).

Soil, Water and Salinity

Chapter III

Salinity of Soils—Effects of Salinity on the Physics and Chemistry of Soils

I. Shainberg

A. Introduction

In areas of low rainfall, salts formed during the weathering of soil minerals are not fully leached. Under humid conditions the soluble salts originally present in soil materials and those formed by the weathering of minerals are generally carried downward into the ground water and are transported ultimately to the oceans. Saline soils are, therefore, practically non-existent in humid regions. Conversely, saline soils occur in arid regions not only because there is less rainfall available to leach and transport the salts, but also because of the high evaporation rates characteristic of arid climates which tend to further concentrate the salts in soils and in surface waters.

Salinity problems of principal economic importance may also arise when previously non-saline soil becomes saline as the result of irrigation. Water used for irrigation may contain from 100 to 1000 g of salt per cubic meter of water. Since the annual application of water may amount to 10000 m^3/ha, the annual addition of salt to the soil may be between 1 and 10 tons/ha. In Israel, for example, the salt content of the water in the National Water Carrier is on the average 700 ppm or 700 g/m^3; the annual addition of salt amounts, therefore, to 7 tons/ha. Thus, considerable quantities of soluble salts may be added to irrigated soils over relatively short periods of time.

In the past, water was frequently plentiful and there was a tendency to use it in excess. Consequently salts did not accumulate in the soil profile but were leached down to the ground water table. However, soils which under natural conditions were well drained and non-saline, often did not have adequate drainage for irrigation. In such soils irrigation induced the rise of the ground water table from considerable depths to within a few meters of the soil surface. The rise of the water table to 1.5–2.0 m below the soil surface, enabled the ground water to rise to the soil surface. Under such conditions ground water as well as irrigation water will contribute to the salinization of the soil.

In modern agriculture and under sprinkler irrigation, the tendency is to apply water in the most efficient and economic way. The amount of water applied in each irrigation is equal to the amount of water that was lost by evapotranspiration from the field. Under such conditions, and if no excess water is applied for leaching, salt accumulates in the soil profile.

The soluble salts that accumulate in soils consist principally of various proportions of sodium, calcium, and magnesium cations and chloride and sulfate anions. Potassium, bicarbonate, carbonate and nitrate ions occur in minor quantities. Borates occasionally occur in small amounts but receive considerable attention because of their exceptionally high toxicity to plants.

A detailed description of the occurrence of salt affected soil was presented elsewhere (BERNSTEIN, 1962). The main purpose of this review is to describe the effect of the concentration and composition of salts on the physical and chemical properties of the soil and to try to explain why the effect of salt is as found. Also, some notes on the reclamation of saline soils are given.

B. Classification and Terminology

Salt affected soils are those which contain excessive concentrations of soluble salts and/or exchangeable sodium. Soluble salts produce harmful effects on plants by increasing both the salt content of the soil solution and the degree of saturation of the exchange complex of the soil with exchangeable sodium. Based on these two factors, the U.S. Salinity Laboratory Staff (1954) classified salt affected soils into three categories:

I. Saline Soils

This group includes soils containing soluble salts in quantities sufficient to interfere with the growth of most crop plants but not containing enough exchangeable sodium to alter soil characteristics appreciably. The amount of salt in a soil above which plant growth is affected depends upon the species of the plant, the texture and water capacity of the soil and the composition of the salt. Thus, the critical concentration of the salt in the soil for distinguishing saline from non-saline soil is arbitrary.

KEARNEY and SCHOFIELD (1936) suggested that saline soils are those which contain more than 0.1% salt. The disadvantage of this definition is that it is independent of the soil properties, mainly its water capacity. In sandy soils with low water capacity, the above percentage of salt when dissolved in the soil solution will cause very high osmotic pressure. Conversely, the above concentration of salt in a heavy-textured soil with large water capacity, will cause only a moderate osmotic pressure. Thus, the U.S. Salinity Laboratory Staff (1954) defined a saline soil as one having an electrical conductivity of the saturation extract greater than 4 millimhos/cm (equivalent to approximately 40 meq/l) and an exchangeable sodium percentage (ESP) less than 15. As will be explained later, owing to the presence of excess salts and the absence of significant amounts of exchangeable sodium, saline soils generally are flocculated and, consequently, the water infiltration and permeability are equal to or higher than those of similar non-saline soils. Upon leaching such a soil with excess water, the salts are removed and a non-saline soil is obtained without the need to add various amendments.

II. Non-Saline Alkali Soils (Sodic Soils)

This group includes soils containing exchangeable sodium in a quantity sufficient to interfere with the growth of most crop plants and not containing appreciable quantities of soluble salts. The decision as to the level of exchangeable sodium in the soil which constitutes an excessive degree of saturation is complicated by the fact that there is no sharp change in the properties of the soil as the degree of saturation with exchangeable sodium is increased. Technically a sodic soil is defined as one whose ESP is greater than 15 and in which the conductivity of the saturation extract is less than 4 millimhos/cm. This is an arbitrary and tentative definition and is discussed later in relation to the effect of cationic species and salt concentration on electrical phenomena at the clay water interface.

The exchangeable sodium present in sodic soil may have a marked influence on the physical and chemical properties of the soil. As the proportion of exchangeable sodium increases, the clay particles in the soil tend to disperse. The dispersed colloids may move and block the pores through which the water flows, thus diminishing the hydraulic conductivity of the soil and causing poor aeration. The pH of these soils usually ranges between 8.5 and 10 due to hydrolysis of adsorbed Na in the absence of electrolytes in the soil solution.

III. Saline Alkali Soils

This term is applied to soils for which the conductivity of the saturation extract is greater than 4 millimhos/cm and the ESP is greater than 15. As long as the concentration of salt in the soil solution is high, the properties of these soils are similar to saline soils, the particles are flocculated and the permeability for water is high. In the presence of excess salt, adsorbed sodium does not hydrolyze (KAMIL and SHAINBERG, 1968) and the pH of these soils is usually less than 8.5. As the concentration of the salts in the soil solution is lowered, for example, due to leaching, the properties of these soils may change markedly and become similar to sodic soils: exchangeable sodium hydrolyzes and the pH increases to values above 8.5, the particles disperse, and the permeability, drainage and aeration become poor. Thus, upon leaching these soils of the excess of salts, amendments such as $CaSO_4$, $CaCl_2$, etc. must be added to the leaching solution.

From the description of the groups of soils affected by salts, it is evident that the type of exchangeable ion and the concentration of the soil solution have very marked effects on the physical and chemical properties of the soils. These two factors affect the macroscopic properties of the soils through their effect on the electrical phenomena at the soil particle-water interface. This topic is discussed from a semi-quantitative viewpoint in the following sections.

C. Distribution of Ions at the Soil Particle-Water Interface

I. The Origin of Electrical Charge on Soil Particles

The soil is essentially a dispersed system, that is, a system in which the particles are in a fine state of sub-division or dispersion. The solid particles of the soils are classified according to their size into sand ($>$ 0.02 mm), silt (0.02—0.002 mm) and clay ($<$ 0.002 mm). The texture of a soil (heavy, loamy, sandy, etc.) is determined by the percentage distribution of the various sizes of soil particles. It is evident that the smaller the particles, the bigger their specific surface area (surface area per gram soils). Thus, the clay particles are those with the highest specific surface area, which may amount to as much as 800 m^2/g for montmorillonite clay, which is the dominant clay in arid and semi-arid soils.

The clay particles have a plate shape. The crystals of the clay minerals are made up of elementary alumino-silicate sheets or lamellea stacked one above another in the C-axis direction. For montmorillonite these lamellea are about 9.5 Å thick and for kaolinite they are 7 Å thick. The plate shape of the clay particles contributes to their high specific surface area.

The clay particles usually carry a net negative charge which is neutralized by adsorbed cations. The charge on the clay particles may arise in two ways:

a) Imperfections within the Crystal Lattice. Substitution of trivalent aluminum for quadrivalent silicone in the tetrahedral layer of the alumosilicate sheet, and magnesium or ferrous ions for aluminum in the octahedral layer of the clay particle lead to a negative charge in the lattice. The charge due to this isomorphous substitution is distributed uniformly in the plate-shaped clay particle.

b) The Particle Charge is Created by the Preferential Adsorption of Certain Ions at Specific Sites, on the Particle Surface. At the edges of the plates, the tetrahedral silica sheets and the octahedral alumina sheets are disrupted and primary bonds are broken. These surfaces resemble those of silica and alumina soil particles. The part of the edge surface at which the octahedral sheet is broken may be compared with the surface of an alumina particle; such particles are either positively or negatively charged, depending on the pH of the solution. Similarly, wherever the tetrahedral sheet is broken, the charge is also determined by the pH, as in the silica particles. Hence, under appropriate conditions, the edge surface area may carry a positive charge.

The particle charge is compensated for by the accumulation of an equivalent amount of ions of opposite sign in the liquid immediately surrounding the particle, keeping the whole assembly electroneutral. The particle charge and the equivalent amount of counter ions together form the electric double layer. The total amount of compensating cations may be determined analytically. This amount, expressed in milliequivalents per 100 g of dry soil, is known as the cation exchange capacity (CEC) of the soil. Consideration of the origin of the clay charge shows that the measured CEC depends on the experimental method (e.g., pH, concentration of the solutions used, the types of replacing cations, the method of leaching out the free electrolytes, the method of preparing the soil samples for measurement, etc.). CHAPMAN (1965) has summarized the experimental difficulties.

Calcium and magnesium are the principal cations found in the soil solution and on the exchange complex of normal soils in arid regions. When the soil is irrigated with saline water, excess soluble salts accumulate in these soils and sodium frequently becomes the dominant cation in the soil solution. Under such conditions, a part of the original adsorbed calcium and magnesium is replaced by sodium.

The physical and mechanical properties of the soils (such as dispersion of the particles, infiltration, permeability, soil structure, stability of the aggregates, etc.) are very sensitive to the type of exchangeable ions present. The divalent ions, mainly calcium, are the ions responsible for many of the physical properties typical of a "good" soil. The deleterious effect of adsorbed sodium on the agricultural soil was described above. Moreover, it was emphasized that even 15% sodium in the exchange complex is sufficient to interfere with the growth of most crop plants. Conversely, excess salt prevents the dispersion effect of adsorbed sodium. Our purpose now is to show where the adsorbed ions are located, what is the effect of salt concentration in the soil solution on the ionic distribution near the clay surface, and why the monovalent ions cause dispersion of the clay particles.

II. The Diffuse Double Layer

The diffuse double layer consists of the particle charge and an equivalent amount of counter ions which compensate for the charge on the particle. The counter ions of the double layers are subject to two opposing tendencies: (i) Electrostatic forces attract the cations to the negatively charged clay surface. (ii) At the same time, these ions have a tendency to diffuse away from the surface of the particle — where the concentration of the counter ions is high — into the bulk of the solution where their concentration is low. This situation is similar to that in the earth's atmosphere in which the air molecules are also subject to two opposing tendencies: gravitational attraction to the center of the earth, and diffusion. The result of these two opposing tendencies is the "atmospheric" distribution, where the concentration of the gas molecules is the highest near the earth's surface, and their concentration gradually decreases with increasing distance from the earth's surface. Similar distribution is obtained in the double layer. The concentration of the counter ions near the clay surface is high and it decreases with increasing distance from the surface (Fig. 1).

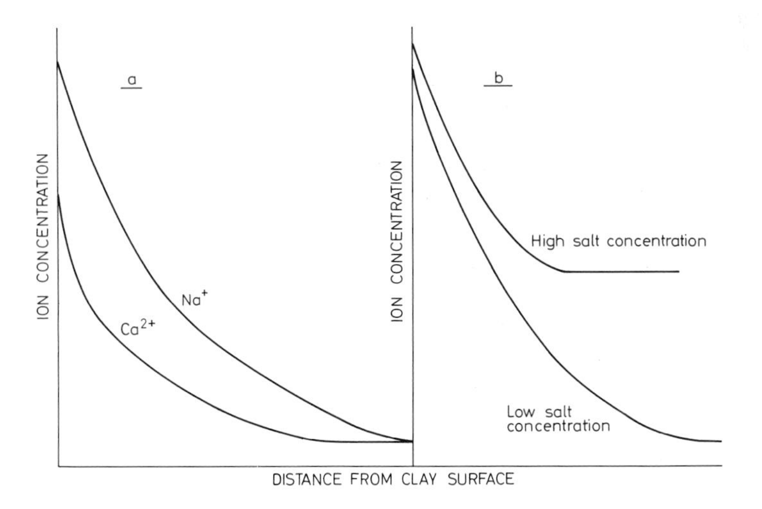

Fig. 1a and b. The distribution of cations near a clay surface (schematic). (a) Mono- and divalent cations (the same salt concentration). (b) Monovalent ions at two electrolyte concentrations

The mathematical description of these two tendencies was presented in detail by OVERBEEK in 1952 (the Gouy-Chapman theory of the diffuse double layer) and is beyond the scope of this review. In this chapter, only the results of the theory are given and explained.

a) Effect of Counter Ion Valency on the Configuration of the Double Layer. Let us compare qualitatively the effect of the two above mentioned tendencies on the configuration of the diffuse double layer, for monovalent (sodium) and divalent (calcium) counter ions systems at the same concentration. The electrical attraction of the surface for the divalent ion is twice the attraction for the monovalent ions. However, the electrolyte concentration in the bulk solution in both systems is the same, so the tendency of the counter ions to diffuse away from the surface is the same in both systems. The net result of equilibrating these two tendencies is that the diffuse double layer in the divalent ion system is more compressed toward the surface. This is presented in Fig. 1a.

b) Effect of Salt Concentration on the Configuration of the Diffuse Double Layer. Let us consider an homoionic system, e.g. a Na-clay system, and an increase in the salt concentration in the bulk solution. Since the only adsorbed ion is sodium, the electrical attraction of the surface for the counter ions is the same, regardless of the bulk concentration. However, when the electrolyte concentration in the bulk solution is greater, the tendency of the counter ions to diffuse away from near the surface to the bulk solution diminishes with an increase in the bulk solution concentration. The net result is that the diffuse counter ion atmosphere is compressed towards the surface when the bulk salt concentration is increased. This phenomenon is presented in Fig. 1b.

D. Swelling and Dispersion of Clay Particles as Affected by Salt-concentration and Ionic Composition

I. Homoionic Na and Ca systems

When two clay particles approach each other, their diffuse counter ion atmospheres begin to overlap. This situation is presented schematically in Fig. 2. It is evident that work must be performed in order to overcome the electrical repulsion forces between the two positively charged ionic atmospheres. The electric double layer repulsion force also called the swelling pressure by soil scientists, can be calculated using the diffuse double layer theory (WARKENTIN et al., 1957). It is evident that the more the ionic atmosphere is compressed toward the clay surface, the less is the overlapping of the atmospheres for a given distance between the particles. Thus, one expects that the repulsive force between the particles decreases with an incrcase in the salt concentration and an increase in the valency of the adsorbed ions. This phenomenon is presented in Fig. 3 where the experimental swelling pressure of Na-montmorillonite is plotted as a function of the moisture retained by the clay particles for 10^{-2} M solution of NaCl. It is evident that the theoretical curve based on the diffuse double layer theory predicts very nicely the experimental swelling pressure of Na-montmorillonite. The calculated and experimental swelling pressures of Ca-montmorillonite in 10^{-3} M $CaCl_2$ solu-

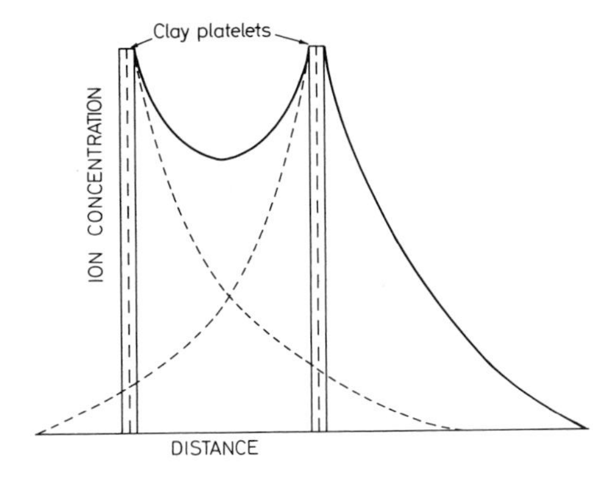

Fig. 2. Overlapping of two diffuse double layers (schematic)

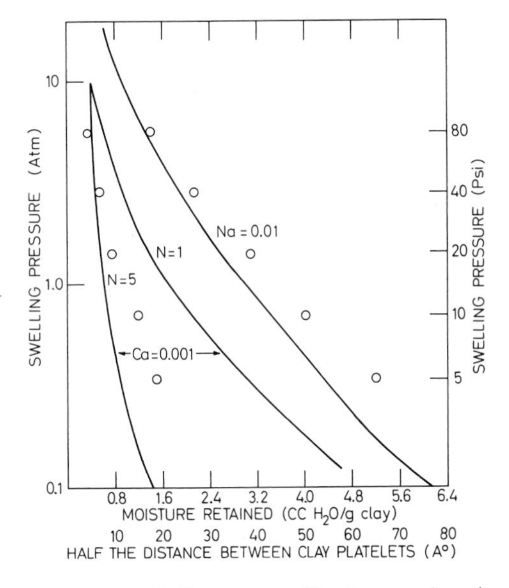

Fig. 3. Swelling pressure of Na- and Ca-montmorillonite as a function of the moisture retained by the clay. (Experimental points and theoretical curves)

tions are also presented. It may be seen that the experimental points for Ca-montmorillonite are far below the predicted Gouy curve.

To appreciate the difference between calculated and experimental swelling of Ca-montmorillonite, it is necessary to consider the arrangement of the clay particles in the gel. The swelling of Na-montmorillonite clay (Warkentin et al., 1957), the negative adsorption of salt by the clay (Bolt and Warkentin, 1958), and the viscosity and light transmission of Na-montmorillonite in suspensions (Shainberg and Otoh, 1968), all indicate that the particles exist as single platelets. Conversely, X-ray diffraction data on Ca-montmorillonite pastes, obtained by Norrish and Quirk (1954) and Blackmore and Miller (1961), indicated that

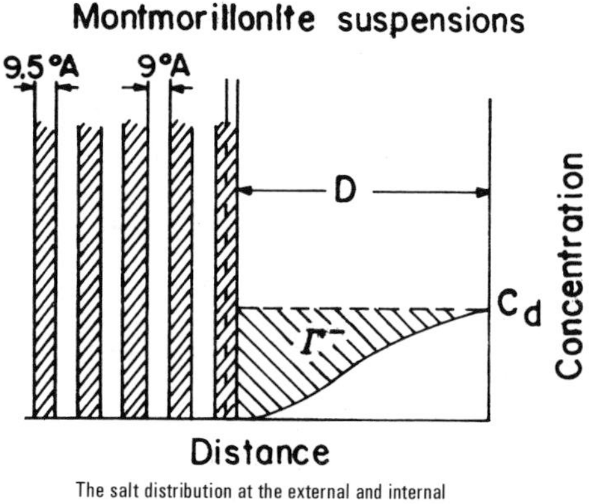

Fig. 4. Schematic presentation of Ca-montmorillonite tactoid and the diffuse double layer at the external surfaces. *D*-distance from external surface of the tactoid to the equilibrium salt solution, where the concentration is C_d. Γ-negative absorption of the salt

Ca-montmorillonite exists in packets or tactoids consisting of several clay platelets each, with a 4.5 Å thick film of water on each internal surface (Fig. 4). It is now accepted that the exchangeable Ca ions adsorbed on the internal surfaces of montmorillonite tactoids do not form a diffuse layer and give only a limited swelling (the C-spacing is fixed at 18.9 Å, even in distilled water). Still, BLACKMORE and MILLER (1961) and SHAINBERG (1968) concluded from swelling pressure and electrochemical measurements, respectively, that a diffuse double layer composed of Ca ions is formed on the external surfaces of the tactoids. This is expressed in Fig. 3, where the experimental swelling points for Ca-montmorillonite appear to fit the theoretical curve calculated for Ca-tactoids, with the number of platelets in a tactoid, N, being 5.

II. Mixed Na/Ca Systems

Saline soils are usually adsorbed with a mixture of Na and Ca ions (although Ca is the dominant ion). Thus, it is important to consider a mixed Na/Ca system. As indicated in the previous section, the platelets in sodium clays are separated, when in equilibrium, with dilute salt solution, whereas calcium-saturated clays exist in tactoids with only a fraction of the particle surfaces active in the swelling-shrinkage cycle. The swelling retention curves for mixtures of mono- and divalent ions can be calculated on the basis of either of the following two models:

a) The adsorbed ions are randomly distributed in the exchange phase.

b) There is "demixing" of the two ions, whereby calcium is concentrated on some interlayers (probably inside the tactoids) and sodium ions are adsorbed on other interlayers.

The volume of water retained by montmorillonite clay saturated with various amounts of exchangeable Na (the complementary ion is calcium) for the various pressures applied is presented in Fig. 5.

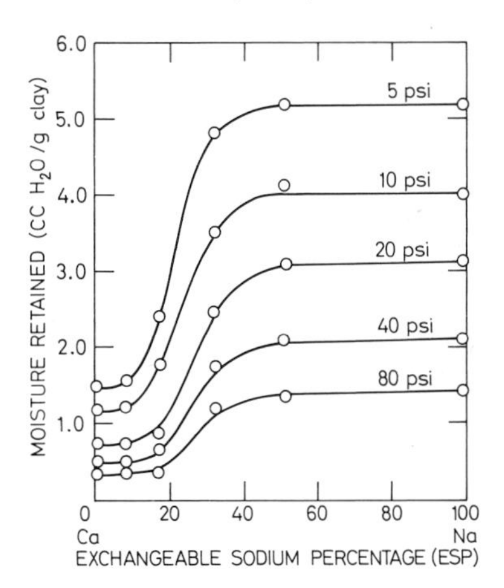

Fig. 5. Water retention by montmorillonite clay as a function of exchangeable sodium percentage (ESP) at various pressures applied

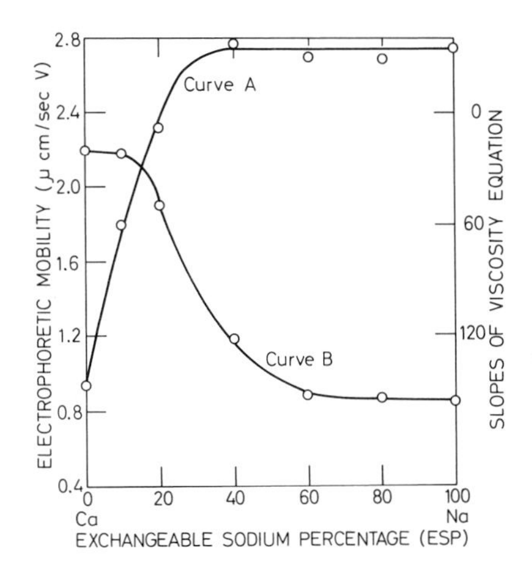

Fig. 6. The dependence of electrophoretic mobility (Curve *A*) and the relative size (Curve *B*) of montmorillonite particles on the exchangeable sodium percentage (ESP). (The relative size is expressed in units of the slope in the Einstein equation for the viscosity of suspension (SHAINBERG and OTOH, 1968)

It is evident from these curves that a slight addition of exchangeable sodium to mainly Ca-saturated clay will have little effect on the amount of solution retained by the clay. Ca-clay and a system with an ESP of 10 will retain about the same volume of solution at each of the pressures applied. The experimental moisture retained as a function of ESP compares favorably with the effect of adsorbed sodium on the size and shape of montmorillonite particles saturated with a mixture of Na and Ca ions in the adsorbed phase (SHAINBERG and OTOH, 1968 and Fig. 6).

These last-mentioned authors found that introduction of a small percentage of Na into the exchange complex of Ca-tactiods was not sufficient to break the packet apart, but that introduction of more Na into the exchange complex (over 20%) did result in the breakdown of the tactoids. The platelets were completely separated when 50–60% of the adsorbed Ca had been replaced by Na. This is in contrast to the effect of sodium on the electrophoretic mobility of montmorillonite particles (BAR-ON et al., 1970, and Fig.6), where a slight addition of exchangeable Na- to Ca-saturated clay has a considerable effect on the electrophoretic mobility of the clay. Replacing 10% of the adsorbed Ca with Na increased the mobility of the particles to twice that of a pure Ca-montmorillonite. When the exchangeable sodium percentage reached a value of about 35%, the electrophoretic mobility of that clay was already identical to that of Na-montmorillonite. Both these phenomena indicate that the basic structure of the Ca-clay particles is not affected by replacing 10% of the adsorbed Ca by Na. Conversely, all of the adsorbed Na concentrated on the external surfaces of the tactoids to give an electrophoretic mobility similar to that of Na-clay. The retention capacity of the clay is more sensitive to the area of the osmotically active surface than to the type of the cation on this surface, thus, the swelling is not affected by introducing 10% Na. Conversely, the dispersion of the tactoids, which depends on the zeta potential (which is the electric potential in the double layer at the interface between a particle which moves in an electric field and the surrounding liquid) and the electrophoretic mobility, is very sensitive to a low percentage of adsorbed sodium.

E. Hydraulic Conductivity of Salt-Affected Soils

The deleterious effect of adsorbed sodium on the physical properties of agricultural soils is most clearly seen in the changes in the permeability of water through the soil. The permeability of the soil is a function of the square of the pore radii (the intrinsic permeability of a soil has the dimensions of cm^2), so that any treatment which decreases the size of the larger pores may have dramatic effects on the permeability of the soil to water. Since the macropores are also those which drain readily after irrigation or rainfall, they also play a significant role in soil aeration and gas exchange.

Swelling of clay particles, by which the size of the large pores is decreased, and dispersion of the soil colloidal material, which can then move and block pores in the soil, may both affect the intrinsic permeability of the soil. It may be deduced from the double layer theory that both swelling and particle dispersion increase as the concentration of salts in the soil solution decreases and as the Na/Ca ratio in the solution increases. Experimental results (see next paragraph) have confirmed that the hydraulic conductivity behaves accordingly, i.e., higher in concentrated solutions and low Na/Ca ratio and lower hydraulic conductivity in dilute solutions and high Na/Ca ratio.

GARDNER et al (1959) have reported that in soil with a low percentage of sodium in the exchange complex, the water permeability of the soil was not more than doubled as the solution concentration was increased from 2 to 100 meq/liter, whereas in high sodium soils permeabilities changed by a few orders of magnitude at similar changes of salt concentration. Data obtained by QUIRK and SCHOFIELD (1955) show similar results for ranges of concentration commonly encountered in soils. The effect of solution concentration on soil permeability in soil with divalent cations was negligibly small.

The quantitative decrease in hydraulic conductivity with decreasing electrolyte concentration and increasing sodium adsorption ratio (SAR) of the percolating solution was assessed by McNeal and Coleman (1968) for seven soils of varying clay mineralogy. They found that the decreases were particularly pronounced for soils high in montmorillonitic clays. Soils high in kaolinite clay minerals and sesquioxides were virtually insensitive to variations in solution composition. It is evident that soils with clay minerals capable of swelling and dispersion are the soils most sensitive to an increase in the SAR of the percolating solution and to a decrease in the electrolyte concentration.

F. Ion Exchange Equilibria

I. Ion Exchange Equations

Ion exchange equations describe the distribution of cations between the adsorbed and solution phases. A number of workers have discussed the merits and limitations of the various exchange equations (Helfferich, 1962; Bolt, 1967; Shainberg, 1972). One of the simplest approaches which leads to a very widely used equation is presented here. The ionic equilibrium and the preference of the clay particles for one of the two counter ions may be described by the mass action law which gives the selectivity coefficient. Thus, for the heterovalent exchange reaction

$$\bar{\mathrm{Ca}} + 2\,\mathrm{Na} = 2\,\bar{\mathrm{Na}} + \mathrm{Ca}\,, \tag{1}$$

where the dashes indicate ions in the adsorbed phase, the selectivity coefficient is given by

$$K_{\mathrm{Ca}}^{\mathrm{Na}} = \frac{(\bar{\mathrm{Na}})^2\,(\mathrm{Ca})}{(\bar{\mathrm{Ca}})\,(\mathrm{Na})^2}\,, \tag{2}$$

where the brackets indicate concentrations of the ionic species. If the concentration of the ionic species in both the adsorbed phase and in solution is given in moles, then Eq. (2) defines the molar selectivity coefficient. More often, the adsorbed ion concentration is expressed in ionic equivalent fraction units, $\bar{x}$, and the concentration in the solution phase in moles per liter, m; thus, Eq. (2) becomes:

$$K = \frac{(\bar{x}_{\mathrm{Na}})^2\,(m_{\mathrm{Ca}})}{(\bar{x}_{\mathrm{Ca}})\,(m_{\mathrm{Na}})^2}\,. \tag{3}$$

The value of the apparent constant in soils can be estimated from the data of Levy and Hillel (1968), who found that over a wide range of equivalent fractions of exchangeable sodium (between 0.1 and 0.7) in typical Israeli soils, the value of this constant was 0.25. Similar observations were made by LAUDELOT et al. (1968) and LEWIS and THOMAS (1963) for the Na–Ca exchange reaction on montmorillonite. This constant is also identical to the "Gapon" constant recommended for use by the U.S. Salinity Staff (1954). Taking the square root of Eq. (3) which means dividing the reaction of Eq. (1) by 2, gives

$$\frac{\bar{x}_{Na}\sqrt{m_{Ca}}}{\sqrt{\bar{x}_{Ca}}\, m_{Na}} = \sqrt{0.25} = 0.5 \tag{4}$$

or

$$\frac{\bar{x}_{Na}}{\sqrt{\bar{x}_{Ca}}} = 0.5\, \frac{m_{Na}}{\sqrt{m_{Ca}}}\,. \tag{5}$$

The Salinity Staff defined a quantity, SAR (Sodium Adsorption Ratio), by

$$\mathrm{SAR} = \frac{m'_{Na}}{\sqrt{m'_{Ca}}} = 31.6\, \frac{m_{Na}}{\sqrt{m_{Ca}}}\,, \tag{6}$$

whereby m' is the solution concentration in mmole/l. Combining Eqs. (5) and (6) gives

$$\frac{\bar{x}_{Na}}{\sqrt{\bar{x}_{Ca}}} = \frac{0.5}{31.6}\,\mathrm{SAR} = 0.0158\,\mathrm{SAR}\,. \tag{7}$$

In the range of a low equivalent fraction of Na in the exchange phase, $\bar{x}_{Ca}$ is close to 1 and $\sqrt{\bar{x}_{Ca}}$ can be replaced by $\bar{x}_{Ca}$; thus, Eq. (7) becomes the familiar equation

$$\mathrm{ESR} = \frac{\bar{x}_{Na}}{\bar{x}_{Ca}} = 0.0158\,\mathrm{SAR}\,. \tag{8}$$

Analyzing a large number of soil samples (The U.S. Salinity Staff, 1954) gave the empirical equation

$$\mathrm{ESR} = -0.0126 + 0.01475\,(\mathrm{SAR})\,. \tag{9}$$

More recently BOWER (1959) published a similar empirical equation in which

$$\mathrm{ESR} = 0.0057 + 0.0173\,(\mathrm{SAR})\,. \tag{10}$$

Since Mg^{2+} does not behave very differently from Ca^{2+}, it is common to combine these two ions together and Eq. (8) becomes

$$\frac{\bar{x}_{Na}}{\bar{x}_{Ca}} = 0.0158\,\frac{(\mathrm{Na})}{\sqrt{(\mathrm{Ca})+(\mathrm{Mg})}}\,, \tag{11}$$

where the parentheses denote the concentration of the ions in the equilibrium solution in mmol/l.

II. The Effect of Total Salt Concentration on the Adsorbed Ion Composition

A key feature of Eqs. (8) or (11) is that to maintain a constant composition of the surface phase the ratio $\frac{(\mathrm{Na})}{\sqrt{(\mathrm{Ca})}}$ must be kept constant. However, by increasing the total concentration of the solution by a factor of two, the numerator is doubled but the denominator is increased only by the square root of two; thus, the SAR of the solution is multiplied by a factor of $\sqrt{2}$ and more sodium is adsorbed on the surface.

The effect of the bulk solution concentration on the composition of the exchanged phase can be derived generally as follows. If, in Eq. (5) instead of the molar concentration, we use the mole fraction units, N, in solution, namely

$$N_{\mathrm{Na}} = \frac{m_{\mathrm{Na}}}{m_0} \quad \text{and} \quad N_{\mathrm{Ca}} = \frac{m_{\mathrm{Ca}}}{m_0} \tag{12}$$

where m_0 is the total molar concentration in solution, then equation (5) gives

$$\frac{\bar{x}_{\mathrm{Na}}}{\sqrt{\bar{x}_{\mathrm{Ca}}}} = 0.5 \frac{N_{\mathrm{Na}} \sqrt{m_0}}{\sqrt{N_{\mathrm{Ca}}}} . \tag{13}$$

For a soil in equilibrium with a solution of $N_{\mathrm{Na}} = N_{\mathrm{Ca}} = 0.5$, the ionic composition of the exchanger is

$$\frac{\bar{x}_{\mathrm{Na}}}{\sqrt{\bar{x}_{\mathrm{Ca}}}} = (0.5)(\sqrt{0.5})\sqrt{m_0} = 0.354 \sqrt{m_0} . \tag{14}$$

Where the solution concentration is $m_0 = 0.01$ m, then

$$\frac{\bar{x}_{\mathrm{Na}}}{\sqrt{\bar{x}_{\mathrm{Ca}}}} = 0.0354 .$$

Since $\bar{x}_{\mathrm{Na}} + \bar{x}_{\mathrm{Ca}} = 1.0$, by solving for $\bar{x}_{\mathrm{Na}}$ one obtains

$$\bar{x}_{\mathrm{Na}} = 0.035 \; \bar{x}_{\mathrm{Ca}} = 0.965$$

and the affinity of the clay for the divalent ion is clearly demonstrated.

The effect of the solution concentration may be demonstrated by substituting the values $m_0 = 1.0, 0.1, 0.01$, and $0.001 \, m$ in Eq. (14). The results are summarized in Table 1. It is evident from Table 1 that the clay prefers the counter ion of higher valence and that the preference increases with dilution of the solution. This effect is readily explained by the diffuse layer potential. The electrical potential attracts the counter ions in a force which is proportional to the ionic charge. Hence, the divalent ion is more strongly attracted and is preferred by the ion exchanger. The absolute value of the electrical potentials increases with dilution of the solution, thus increasing the affinity of the clay for the divalent ion, with increasing dilution.

Table 1. The effect of solution concentration on the equivalent fraction of adsorbed cations from solutions of a constant mole fraction. ($N_{Na} = 0.5$ $N_{Ca} = 0.5$)

Solution concentration (mole/l)	Equivalent fraction of adsorbed ion	
	Na	Ca
1.0	0.353	0.647
0.1	0.106	0.894
0.01	0.035	0.965
0.001	0.011	0.989

G. Reclamation of Saline Soils

Saline soil has been defined as soil containing enough soluble salts so distributed in the soil profile as to decrease the growth of most plants (REEVE and FIREMAN, 1967). The reclamation process is aimed at removing this salinity-limiting factor so that the crop yield will not suffer any reduction.

I. Reclamation of Initially Saline Soils by Leaching

Reclamation of initially saline soil is essentially a process in which the high concentration soil solution is displaced by a less concentrated solution. Miscible displacement concepts have been applied to the reclamation of saline soils by removal of soluble salt from them (BIGGAR and NIELSEN, 1967). The simplest flow model is the piston flow, namely, that there is no mixing at the boundary between the water flowing into the soil and the water present in the soil. Although such flow seldom, if ever, takes place, the assumption can be useful for calculating the salinity profile at any given time.

A more realistic approach is to consider the mixing that takes place at the boundary between the leaching and the soil solutions. The mixing of salt at the boundary is due to three processes:

a) Hydrodynamic dispersion. The porous soil consists of small and large capillaries in a normal distribution. The displacing solution flows faster in the larger pores, thereby resulting in a mixing of salt at the boundary.

b) Diffusion. Diffusion of ions and salts from the concentrated soil solution to the displacing solution takes place at the boundary.

c) Adsorption and ion exchange reaction. If an adsorption reaction takes place between the soil particles and the ions in the displaced and displacing solution, the boundary between the two solutions is much more diffused.

These three effects were combined by BRESLER (1972) to describe the dynamics of salts in the soil. The theory suggests and field experiments confirmed that reclaiming saline soil by leaching is more efficient when the soil is maintained unsaturated and the flow rate is relatively slow.

II. Salinity Control during Irrigation

The amount of salt added by irrigation is determined by the quality and the amount of the irrigation water used. Evapotranspiration of the water leaves the salt to accumulate in the root zone unless means are provided for its removal. The most common method by which salt is removed is by leaching it out of the root zone with water.

The average salt concentration of the root zone does not change appreciably if the volume of irrigation water D_i (the volume of water per unit area of land is expressed by its depth), times its salt concentration, C_i is equal to the volume of water draining out of the soil, D_d times the salt concentration of this drainage water, C_d. An equation for this is

$$D_i C_i - D_d C_d = 0 . \tag{15}$$

The concept of leaching requirement (LR) is now naturally derived from Eq. (15) and is given by Eq. (16)

$$LR = \frac{D_d}{D_i} = \frac{C_i}{C_d} . \tag{16}$$

It is evident from Eq. (16) that, as the salinity of the irrigation water increases, so does the leaching requirement, namely, a higher percentage of irrigation water must be drained in order to prevent salinization of the soil.

Good crop yields are dependent on the maintainance of the salt concentration of the soil solution in the root zone at or below certain levels. The levels required for many crops have been determined and published (U.S. Salinity Laboratory Staff, 1954) in terms of the electrical conductivity of saturation extracts, which is about twice that of a soil solution at field capacity. If the movement of soil solution down through the soil profile were perfectly uniform and steady, the concentration of soil solution at the bottom of the zone from which the roots extract solution would be practically equal to the concentration of the drainage water. Consequently, if one of the objectives of irrigation is to keep the salt concentration below the levels deleterious to crops, the concentration of the drainage waters, C_d, must be at most equal to the critical concentration typical to the specific crop. Knowing the concentration of salts in the irrigation water, C_i, and the critical concentration of the crop, C_d, the percentage of irrigation water that must be drained is easily calculated from Eq. (16).

Rainfall is not included in the leaching requirement equation as presented above. When rainfall is evenly distributed throughout the year or the irrigation season, the rainfall may be considered by taking a weighted average salt concentration, C_{ir}, as follows

$$C_{(ir)} = \frac{C_i D_i + C_r D_r}{D_i + D_r} , \tag{17}$$

where D_r and C_r are the depth and salt concentration of the rainwater, respectively.

However, one has to remember that in applying Eqs. (15)–(17), a steady state restriction was imposed on the systems by the assumptions made. Thus, when the rain falls during a short rainy season, as is the case in the Mediterranean climatic region, this approach might result in erroneous conclusions. This was demonstrated by SHALHEVET (1973) who showed that in light to medium soils in regions of 500 mm rainfall, the leaching of salt during the rainy season is very efficient and complete, and thus these soils will not require any leaching at all during the irrigation seasons.

H. Conclusions

Salt affected soils are soils which contain excessive concentrations of either soluble salts or exchangeable sodium, or both. It was emphasized that saline soils had good physical and chemical properties, e.g., high permeability for water, no dispersion, good aeration, neutral pH, etc. However, the accumulation of salt produces harmful effects on plants by increasing the osmotic potential of the soil solution. Thus, salt must be removed in order to have a successful crop. Whenever there is a crop which can tolerate high salt concentration in the soil solution, there is no need to leach out the salt from the soil profile.

Conversely, exchangeable sodium has deleterious effect on the physical and chemical properties of the soil. Alkali soils have high pH, are dispersed, have poor aeration, etc. Thus, in order to have a successful crop, the sodic soil must be reclaimed. Whereas reclamation of saline soils is relatively easy (leaching the soil), the reclamation of alkali soils is more difficult; water does not penetrate and amendments like Ca-salts must be added in order to replace the exchangeable sodium.

References

BAR-ON, P., SHAINBERG, I., MICHAELI, I.: The electrophoretic mobility of Na/Ca montmorillonite particles. J. Colloid Interface Sci. **33,** 471–472 (1970).

BERNSTEIN, L.: Salt affected soils and plants. UNESCO Arid Zone Res. **18,** 139–174 (1962).

BIGGAR, J.W., NIELSEN, D.R.: Miscible displacement and leaching phenomena. In: Irrigation of agricultural lands, pp. 254–274. Agronomy 11. Madison, Wisc.: Am. Soc. Agronomy 1967.

BLACKMORE, A.V., MILLER, R.D.: Tactoid size and osmotic swelling in Ca montmorillonite. Proc. Soil Sci. Soc. Am. **25,** 169–173 (1961).

BOLT, G.H.: Cation exchange equations used in soil science — a review. Neth. J. Agr. Res. **15,** 81–103 (1967).

BOLT, G.H., WARKENTIN, B.P.: The negative adsorption of anions by clay suspensions. Kolloid-Z. **156,** 41–46 (1958).

BOWER, C.A.: Cation exchange equilibria in soils affected by Na salts. Soil Sci. **88,** 32–35 (1959).

BRESLER, E.: Control of soil salinity. In: HILLEL, D. (Ed.): Optimizing the soil physical environment toward greater crop yields, pp. 101–139. New York: Academic Press 1972.

CHAPMAN, H.D.: In: BLACK, C.A. (Ed.): Cation exchange capacity in Methods of soils analysis, Part 2, pp. 891–900 (1965).

GARDNER, W. R., MAYHOUGH, M. S., GOERTZEN, J. Q., BOWER, C. A.: Effect of electrolyte concentration and exchangeable sodium percentage on diffusivity of water in soils. Soil Sci. **88,** 270–274 (1959).

HELFFERICH, F.: Ion exchange, Chapter 5. New York: McGraw-Hill 1962.

KAMIL, J. SHAINBERG, I.: Hydrolysis of sodium montmorillonite in NaCl solutions. Soil Sci. **106,** 193–199 (1968).

KEARNEY, T. H., SCHOFIELD, C. S.: The choice of crops for saline land. U.S. Dept. Agr. Circ. **404** (1936).

KELLEY, W. R.: Alkali soils. New York: Reinhold Publ. Corp. 1951.

LAGERWERFF, J. V., NAKAYAMA, F. S., FRERE, M. H.: Hydraulic conductivity related to porosity and swelling of soil. Proc. Soil Sci. Soc. Am. **33,** 3–11 (1969).

LAUDELOUT, H., VAN BLADEL, R., BOLT, G. H., PAGE, A. L.: Thermodynamics of heterovalent cation exchange reactions in a montmorillonite clay. Trans. Faraday Soc. **64,** 1477–1488 (1968).

LEVY, R., HILLEL, D.: Thermodynamics equilibrium constants of Na/Ca exchange in some Israeli soils. Soil. Sci. **106,** 393–398 (1968).

LEWIS, R. J., THOMAS, H. C.: Adsorption studies on clay minerals. VIII. A consistency test of exchange sorption in the systems Na/Ca/Ba montmorillonite. J. Phys. Chem. **67,** 1781–1783 (1963).

MCNEAL, B. L., COLEMAN, N. T.: Effect of solution composition on soil hydraulic conductivity. Proc. Soil Sci. Soc. Am. **32,** 308–312 (1968).

MCNEAL, B. L., NORVELL, W. A., COLEMAN, N. T.: Effect of solution composition on soil hydraulic conductivity and on the swelling of extracted soil clays. Proc. Soil Sci. Soc. Am. **30,** 308–317 (1966).

NORRISH, J., QUIRK, J. P.: Crystalline swelling of montmorillonite. Nature **173,** 255–256 (1954).

OVERBEEK, J. TH. G.: In: KRUYT, H. R. (Ed.): Colloid science, Chapters 4–6, Vol. I. Amsterdam: Elsevier 1952.

QUIRK, J. P., SCHOFIELD, R. K.: The effect of electrolyte concentration on soil permeability. J. Soil Sci. **6,** 163–178 (1955).

REEVE, R. C., FIREMAN, M.: Salt problems in relation to irrigation. In: Irrigation of agricultural lands, pp. 988–1008. Agronomy 11. Madison, Wisc.: Amer. Soc. Agronomy 1967.

SHAINBERG, I.: Electrochemical properties of Na and Ca montmorillonite suspensions. Trans. 9th Int. Congr. Soil Sci. **1,** 577 (1968).

SHAINBERG, I.: Cation and anion exchange reactions. In: Chesters, Brenner (Eds.): Soil chemistry. New York: Marcel Decker Co. 1973.

SHAINBERG, I., OTOH, H.: Size and shape of montmorillonite particles saturated with Na/Ca ions. Israel J. Chem. (1968).

SHALHEVET, J.: Irrigation with saline water. In. YARON, B., DANFORS, E., VAADIA, Y. (Eds.): Arid zone irrigation, Ecol. Studies, Vol. 5, Berlin-Heidelberg-New York: Springer 1973.

U.S. Salinity Laboratory Staff.: Saline and alkali soils. Agriculture Handbook No. 60, U.S.D.A. (1954).

WARKENTIN, R. P., BOLT, G. H., MILLER, R. D.: Swelling pressures of Montmorillonite. Proc. Soil. Sci. Soc. Am. **21,** 495–499 (1957).

Chapter IV

Water Quality for Irrigated Agriculture

L. D. Doneen

A. Introduction

Irrigation is generally practiced in the arid regions of the world, where evapotranspiration is greater than the precipitation. The waters of these areas have a greater tendency to be of poor quality than those in the humid regions. The earth crust in its weathering, disintegration and decomposition in the process of soil formation liberates many soluble salts. This process is greatly enhanced by the high concentration of carbon dioxide, a weak acid, in the soil atmosphere and soil solution along with the humic acids from decomposition of organic matter. Where rainfall is sufficiently high to leach the soil with good drainage, the salts are carried away by the rivers, or by seepage to the ocean, but in the arid lands where the soils are not highly leached and surface drainage is poor the salts have a tendency to accumulate.

In general, the ground waters contain more salts than the replenishing surface water. This is due to several factors but the principal ones are: a) leaching of salts applied in the surface water by occasional heavy rain or excess irrigation after being concentrated in the surface layers of soil by evapotranspiration, and b) by dissolving of minerals by the water percolating from the soil mantle above the ground water basin. If the soil mantle contains lime or gypsum, this will greatly enhance the salt content of the percolating water.

A number of criteria have been devised for the classification of water quality for irrigation. These schemes vary from general to detailed classifications for a particular crop or region. In addition to the chemical analysis of the water many other factors require evaluation, such as soil properties, irrigation management, climate and crops, before determining its suitability for irrigation.

Most water quality classifications have been developed for an arid climate where the annual rainfall rarely wets the soil to two meters in depth. Consequently most of these standards may not be applicable to a monsoon climate or to areas where high precipitation occurs part of the year followed by long periods of drought.

This chapter will trace the historical development of water quality standards, the current knowledge on the subject and factors that may limit or nullify their use under certain conditions. A complete literature review is not intended, but only those publications making an important contribution to the subject. The California State Water Quality Control Board (1952) reviewed the criteria for all ordinary uses of water and includes a very extensive bibliography.

B. Composition of Irrigation Waters and Method of Reporting

All irrigation water contains dissolved salts in varying amounts. The total concentration and the important constituents determine the quality of the water. When salts go into solution they separate into ions-cations and anions. The principal cations of irrigation water are; calcium (Ca), magnesium (Mg), sodium (Na), and potassium (K) which usually is not determined because of its low concentration in many waters. The anions are carbonate (CO_3), bicarbonate (HCO_3), chloride (Cl), sulfate (SO_4), and nitrate (NO_3) which normally is not determined because of its low concentration, but there are exceptions. In addition boron (B), should be included for those areas where it occurs. The above ions and other minor constituents of less importance to the quality of irrigation water are given in considerable detail by WILCOX (1948) and by HEM (1959).

The quality of an irrigation water is judged not only by the total, but by the kind of salts it contains and the individual ions involved. Ions have been reported in several different units, but by general consensus, most analyses are now reported as milligram equivalents (me) or milliequivalents per liter me/l (these are identical terms), or as milligrams per liter (mg/l). For the following presentation the commonly accepted term for agricultural water of milliequivalents per liter will be used, which allows us to work with chemical equivalents or combining weights of salt constituents of the water and cation exchange of the soil. To change from milligrams per liter to me/l divide the mg/l by the equivalent weight of the ion or salt given in Table 1.

Table 1. Equivalent weights of ions in irrigation water

Cations	Equivalent weights	Anions	Equivalent weights
Calcium (Ca)	20.0	Carbonate (CO_3)	30.0
Magnesium (Mg)	12.2	Bicarbonate (HCO_3)	61.0
Sodium (Na)	23.0	Sulfate (SO_4)	48.0
Potassium (K)	39.1	Chloride (Cl)	35.5

Water analyses published before 1900 are practically all expressed in terms of concentrations of combined salts such as calcium sulfate, sodium chloride, or oxide of the cations as CaO or MgO. In the USA, parts per million (ppm) have been in common use for certain analyses which is the same as mg/l, however, the USA is gradually shifting to the metric system. Some older analyses are reported in terms of parts per 100000. This is no longer in general use. In the English system of units analyses are sometimes expressed in grains per gallon. To convert grains/gal to mg/l, multiply by 17.1 for the US gallon and by 14.3 for the British gallon.

In the older analyses the total salt concentration was reported as total dissolved solids (TDS) in ppm by evaporating to dryness an aliquot of water and weighing the dry residue. In general, this method has been replaced by electrical conductivity *(EC)* measurement which is the reciprocal ohms or mhos/cm as a standard unit, at 25 °C. Natural waters have specific conductance considerably less than one mho, therefore, millimhos/cm (mmhos/cm) or micromhos/cm (μmho/cm) are used in estimating the salt concentration of water as illustrated in Table 2 for Colorado River water. In the USA conductive measurements of $K \times 10^5$ (equal to $EC \times 10^5$) were in general use until about mid-1950's when a gradual shift to the use of $EC \times 10^3$ or $EC \times 10^6$ occurred. The U.S. Salinity Laboratory (1954) suggested the use of $EC \times 10^6$ for irrigation water and $EC \times 10^3$ for the more concentrated soil solution extracts. However, many agencies use $EC \times 10^3$ for both water and soil extracts. HEM (1959) reviews other minor methods of reporting and presenting water analyses. Electrical conductance is an excellent and rapid method for obtaining an estimate of total salt content, within an accuracy of about 10 percent for most irrigation waters. However, individual salts that may predominate are not indicated and analyses for these must be made for interpreting the quality of an irrigation water.

Table 2. Electrical conductivity expressions and factors for estimating me/l and mg/l

Electrical conductivity	Est. me/l Multiply EC by	Est. mg/l Multiply EC by
EC = 0.00117 mho/cm		
$EC \times 10^3$ = 1.17 millimhos/cm	10 = 11.7	640 = 749
$EC \times 10^6$ = 1170 micromhos/cm	0.01 = 11.7	0.64 = 749
$K \times 10^5$ [a] = 117 or $EC \times 10^5$	0.1 = 11.7	6.4 = 749

[a] The electrical conductivity of $K \times 10^5$ was used in the USA until about mid-1950's.

Sodium has been long recognized as having an adverse effect on the soil when the ratio of Na to Ca and Mg is high. This is expressed by many water classifications as Na percentage according to the following equation; the cations are expressed in me/l:

$$\text{Na percentage} = (\text{Na} \times 100)/(\text{Ca} + \text{Mg} + \text{Na}).$$

Boron because of its low concentration is expressed as mg/l or ppm of the element and is often not determined unless in an area where boron toxicity is suspected. When work is reviewed in this chapter the analyses will be reported in terms used in the original publication.

C. Early Water Quality Investigations

Research on the quality of irrigation water dates back only a few decades. However, during the latter part of the nineteenth century Hilgard's pioneering work on water quality showed the importance of composition and he rated water by the anion content as well as by the total salt concentration (HILGARD, 1906). Based on

Hilgard's work, STABLER (1911) rated sodium carbonate as being twice as undesirable as sodium chloride and 10 times as undesirable as Na_2SO_4 for irrigation waters, which indicated the importance of the bicarbonate ion in evaluating water quality for irrigation, but was ignored for about 50 years. As the knowledge of cation exchange developed, the role of sodium in dispersing the soil, reducing it permeability and the development of sodic conditions, resulted in the determination of cations playing a major role in the evaluation of irrigation water quality.

In 1931 a group of scientists in California published, by mimeograph leaflet, tentative standards as a guide to the quality of irrigation water. DONEEN (1954) traces the evolution of this original draft through a series of changes until it emanated in 1943 in the form given in Table 3. This water quality classification, in its earlier versions or modifications, has been published by many authors and has wide distribution throughout the western USA. The authors recognized that owing to diverse climatological conditions, crops and soils, it is not possible to establish rigid limits for all conditions. The three broad classes listed in Table 3 should be interpreted as follows:

Class 1. Excellent to Good—Regarded as safe and suitable for most plants under any condition of soil or climate.

Class 2. Good to Injurious—Regarded as possibly harmful for certain crops under certain conditions of soil or climate, particularly in the higher ranges of this class.

Class 3. Injurious to Unsatisfactory—Regarded as probably harmful to most crops and unsatisfactory for all but the most tolerant.

Table 3. Qualitative classification of irrigation waters

Item	Class 1 Excellent to good	Class 2 Good to injurious	Class 3 Injurious to unsatisfactory
$K \times 10^{5}$ [a] at 25° Ca	Less than 100	100 —300	More than 300
Boron, ppm	Less than 0.5	0.5— 2.0	More than 2.0
Sodium, pct [b]	Less than 60	60 — 75	More than 75
Chloride, me/l	Less than 5	5 — 10	More than 10

[a] $K \times 10^5 = EC \times 10^5$.

[b] Sodium percentage $= \frac{Na \times 100}{Ca + Mg + Na}$ Ions in me/l.

Only one item in Table 3 is required to place the water in a higher class. This same type of interpretation applies to all water classifications to be presented in this paper regardless as to whether three, or a large number, of classes are involved. The above classification was based primarily on field experience and related studies in soils and agronomic practices with little work on water quality *per se*.

D. Recent Studies on Water Quality

I. Wilcox Water Classification

WILCOX (1948) published a diagram showing five classes of water (Fig. 1). This classification is based essentially on the earlier work of SCOFIELD (1936) which also has the same number of classes based on total salt content and percent sodium, and 15 classes for boron according to the concentration in the water and crop sensitivity to boron. This classification, particularly the diagram, was rather popular for several years, and appears sometimes with some variations, in several publications on the evaluation of irrigation waters. Wilcox's classification appears to be a forerunner to the U.S. Salinity Laboratory classification as many waters fall into the same general class for both criteria.

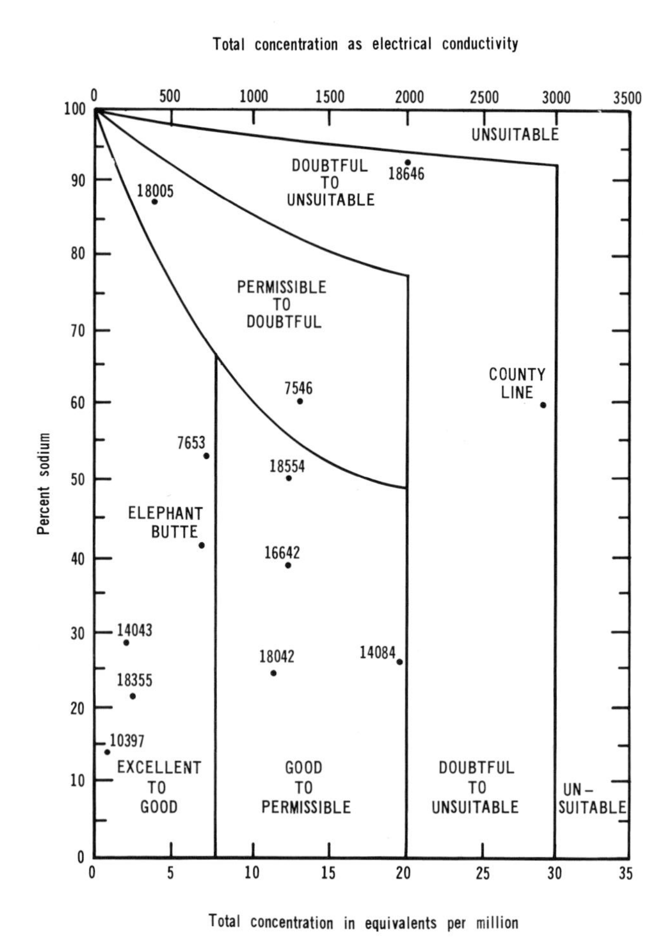

Fig. 1. Diagram for use in interpreting the analysis of an irrigation water. (WILCOX, 1948. A forerunner to U.S. Salinity Laboratory classification; not in general use)

II. U.S. Salinity Laboratory Water Classification

This classification has received world-wide distribution by Handbook No. 60 (U.S.S.L. 1954) and has been reproduced many times in publications concerned with irrigation and water management as in the widely circulated publications: Irrigation of Agricultural Lands (1967), International Source Book on Irrigation and Drainage of Arid Land (1973) and Physical Edaphology (1972). The essence of the Salinity Laboratory's classification is presented in diagram form (Fig. 2). It is based on the interaction of total salt concentration and sodium concentration which is expressed as the sodium adsorption ratio SAR and is defined as

$$\mathrm{SAR} = \frac{\mathrm{Na}}{\sqrt{\mathrm{Ca} + \mathrm{Mg}/2}} \quad \text{Ions in me/l.}$$

At equilibrium the SAR is closely related to the exchangeable sodium of the soil.

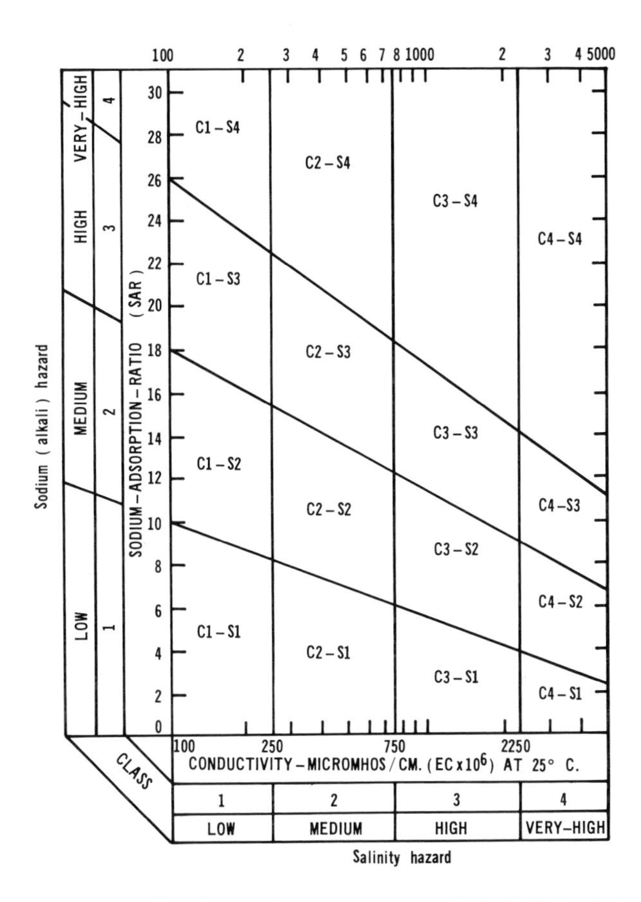

Fig. 2. Diagram for the classification of irrigation waters. (U.S. Salinity Laboratory, 1954, with world wide distribution, but does not apply to a large group of waters)

With time the validity of this classification has been questioned for a large group of waters. The author working under field conditions with many different water qualities found about as many disagreements with this system of classification as agreements, and concluded that it was no better than the one given in Table 3. A conference of scientists on water criteria, (LUNT, 1963) expressed considerable criticism of the Salinity Laboratory's classification. It was recommended that it be modified and that additional research be carried out for its improvement because it appears to be excessively conservative and overestimates the salinity hazard. Furthermore, the SAR appears to be poorly correlated with soil permeability in relation to the sodium status of the soil.

III. Role of Carbonates in Irrigation Water

EATON'S (1949, 1949a) investigations revived the earlier concept that bicarbonate is an important constituent in the evaluation of irrigation water quality. With the concentration of the soil water by evapotranspiration CO_3 and HCO_3 precipitate as Ca-$MgCO_3$. With the loss of Ca and Mg from the water, the relative proportion of Na is increased, resulting in an increased sodium hazard. When the carbonate-bicarbonate exceeds the Ca and Mg EATON designated this as residual sodium carbonate, (RSC) and he defined it as; $RSC = (CO_3 + HCO_3) - (Ca + Mg)$ in me/l. If the concentration of Ca and Mg is greater than the HCO_3 precipitation will occur, resulting in an increased precent Na of the soil water, which EATON designated as:

$$\text{percent sodium possible} = \frac{Na \times 100}{(Ca + Mg + Na) - (CO_3 + HCO_3)} \quad \text{Ions in me/l.}$$

These reactions will increase the sodium hazard and indicate an additional parameter for water quality standards. EATON did not set standards, but indicated that increasing amounts of RSC in the irrigation water would hasten the development of sodic soils. However, WILCOX et al. (1954) growing plants in a pot experiment irrigated with a series of HCO_3 and Cl waters, concluded that waters containing more than 2.5 me/l of RSC are not suitable for irrigation, those containing between 1.25 and 2.5 me/l are marginal, and those of less than 1.25 me/l are probably safe. Several investigators of water quality have adopted these standards as an addendum to one of the water quality standards.

The author (unpublished) conducted a similar experiment to the one listed above, using waters of 15 me/l with various concentrations of Ca-Na with HCO_3—SO_4 and obtained essentially the same result as WILCOX; namely, much of the HCO_3 in the water was precipitated in the soil with a corresponding increase in exchangeable-sodium percentage (ESP) and infiltration was not drastically reduced until the irrigation water contained 5 me/l or more of RSC. However, high levels of ESP were obtained at the lower levels of RSC which was contrary, at that time, to the acknowledged role of Na in soil permeability.

IV. Preliminary Research to Doneen's Water Classification

For over a 20-year period, DONEEN and co-workers conducted many experiments on water quality involving the field, greenhouse and laboratory. Much of this work stemmed from water quality problems encountered under field conditions and research results from greenhouse and laboratory were verified in irrigated areas using waters of varying qualities for irrigation. From this mass of data—published and unpublished—emerged DONEEN's water quality classification and for a better interpretation the research back-up for this classification is reviewed.

a) Salinity. The salt concentrations present in most irrigation waters are not sufficiently high to be injurious to plant growth. However, waters having appreciable salt concentration contribute to soil salinity, thus requiring an excess amount of irrigation water to leach the accumulated salts below the root zone. Plants absorb water and release it by transpiration; some water is lost by surface evaporation. These processes result in most of the salt in the irrigation water remaining in the soil. Usually a good productive soil will have a soil solution 4–10 times more concentrated than the water applied. If the water is used sparingly or a high water table develops which prevents downward percolation of excess water, the concentration of the soil solution may be 40–80 times that of the original water. Thus the soil becomes too saline for plant growth. Some drainage or leaching is required to maintain an irrigated agriculture.

Salts normally found in natural waters have a wide range of solubility and may be roughly classified into low and high solubility, shown in Table 4.

A concentration of 80 me/l in the soil solution at field capacity is usually considered to be indicative of incipient salinity. When extracted at saturation the lower limits of salinity are usually 40–50 me/l. Therefore, the salts of low solubility such as $CaCO_3$, $Ca\text{-}MgCO_3$ or $CaSO_4$ precipitate before they reach a saline concentration in the soil solution.

Table 4. Solubility of salts in milliequivalents per liter of water

Low solubility	me/l	High solubility	me/l
Calcium carbonate ($CaCO_3$)	0.5[a]	Calcium chloride ($CaCl_2 \cdot 6H_2O$)	25470
Calcium bicarbonate $Ca(HCO_3)_2$	3–12[a]	Magnesium sulfate ($MgSO_4 \cdot 7H_2O$)	5760
Calcium sulfate ($CaSO_4 \cdot 2H_2O$)	30	Magnesium chloride ($MgCl_2 \cdot 6H_2O$)	14955
Magnesium carbonate ($MgCO_3$)	2.5	Sodium bicarbonate ($NaHCO_3$)	1642
Magnesium bicarbonate ($Mg(HCO_3)_2$)	15–20[a]	Sodium sulfate ($Na_2SO_4 \cdot 10H_2O$)	683
		Sodium chloride (NaCl)	6108

[a] Solubility will be influenced by the concentration of carbon dioxide (CO_2 in the solution and soil air).

Many waters, particularly in arid regions, are partly or nearly saturated with $Ca(HCO_3)_2$ which upon concentration should precipitate in the soil. DONEEN (1959) investigated this possibility in a citrus growing area of California where high yielding orange and lemon groves were irrigated with water ranging in concentration from 11–37 me/l. These crops, being salt sensitive, should not thrive, or grow at all, according to the water classifications in general use.

By assuming that Ca and $MgCO_3$ and $CaSO_4$ precipitated (Table 4) before a saline condition was reached in the soil, the average salinity of the waters was calculated to be reduced by half. The highly soluble salts remaining in the soil water were designated as constituting the "effective salinity". The above investigation was conducted in 1952 and from other research on the precipitation of salts of low solubility DONEEN (1954) proposed a tentative classification for the "effective salinity" of irrigation waters. Later work by DONEEN (1958), DONEEN and HENDERSON (1960), and TANJI and DONEEN (1966) verified that salts of limited solubility do not create a saline condition in the soil.

b) The Sodium and Bicarbonate Relationship. With increasing concentration of the soil water $Ca(HCO_3)_2$ will precipitate while $Mg(HCO_3)_2$ being more soluble will replace the Ca on the exchange complex. Thus, removing Ca and Mg from the soil water will greatly increase the Na ratio of the soil solution with a corresponding increase in ESP, as illustrated by DONEEN (1958) and DONEEN and HENDERSON (1960). There will be a marked reduction in the infiltration rates of the irrigation waters. This reduction in soil permeability is closely related to the amount of residual $NaHCO_3$ in the water.

c) Salt Concentration in Irrigation Water and Soil Permeability. Until recently the effect of the salt concentration of the irrigation water as affecting permeability has received little attention. This has been the case even though it has been known for a long time that saline-alkali soils have extremely low permeabilities after the salts are removed by leaching. FIREMEN and BODMAN (1939) and FIREMAN (1944) have shown a wide difference in infiltration rates between waters of high and low salt concentrations. Working independently and utilizing different techniques, QUIRK and SCHOFIELD (1955) and HENDERSON (1958) came to the same conclusion, that increasing the salt concentration (in the range for irrigation waters) and holding the *SAR* constant increased the infiltration rate. The infiltration curves plotted for various *SAR* and concentrations of salt in water were hyperbolic with the largest increase in infiltration occurring between 0 and 10 me/l. Even with waters in equilibrium with the soil at ESP of 20, good permeability was obtained with waters in the 20–30 me/l range.

The above laboratory experiments were supplemented in a series of experiments using deep lysimeters. Examples are reports by DONEEN (1958), and DONEEN and HENDERSON (1960). The results were in very close agreement with those of the laboratory experiments. With the deep lysimeters the surface soil tends toward equilibrium with the applied water and the salts were deposited in the subsoil. Under field conditions the salts were often leached below the depth of crop rooting. These results indicate that high salt concentrations in the irrigation water, with sodium predominating, tend to keep the soil flocculated and maintain a reasonably good permeability. From this series of experiments evolved the "permeability index" for irrigation waters.

A problem which arose in DONEEN's earlier work and probably also in that of WILCOX et al. (1954) was the accumulation of soluble salts on the soil surface by evaporation from the shallow pots used in the experiments. Consequently irrigation water, regardless of its composition, was greatly increased in salt concentration and relatively high infiltration rates resulted. This condition would only occur in the presence of very shallow water tables or in soil where high concentration of soluble salts may accumulate on the soil surface by evaporation between irrigations.

d) Soil Conditions. Attempts have been made to anticipate the effect of water on soils on the basis of the law of mass action. Precipitation and dissolution of salts, cation exchange, leaching and drainage, soil texture and clay mineral types, and other dynamic factors not included in mass action evaluation, profoundly affect the predicted equilibrium. In this context DONEEN's "effective salinity" criteria involved only the constituents in the water and did not consider the Ca that may be derived from the exchange complex, if the irrigation water is deficient in Ca, for precipitation of the HCO_3 and/or the SO_4.

For the first time DONEEN (1954) introduced a soil characteristic, leachability in rating the quality of irrigation waters. The three classes of soil leaching are: (1) little leaching due to low percolation rates; (2) some leaching but restricted-deep percolation, or slow drainage; (3) deep percolation of water easily accomplished in open soils. This classification of soil leaching is not related to soil texture but to conditions affecting soil permeability such as high sodium, stratification, and the presence of clay lenses, some types of clay and adobes, dense or compact subsoils and heavy clay subsoils. In addition, high water table soils may have restricted drainage which prevents leaching of the soil, while others may have, to a varying degree, a natural or artificial drainage system which provides for salt removal. Most hardpans and some clay pans are practically impermeable to water. Even these may have cracks or ruptures through which some water and accumulated salts may drain. Different standards were established for "effective salinity" of irrigation waters under the different regimes of restricted drainage than were defined for deep open soils where leaching is easily accomplished.

V. Doneen's Water Quality Classification

From the above research evolved a modified criterion for estimating the quality of agricultural waters. The first preliminary report was published by DONEEN in 1961, and this was modified in 1967. Salt tolerance of plants was added in 1970 (DONEEN 1961, 1967, 1970). The classification is based on three criteria; potential salinity, permeability index, and toxic substances.

a) Potential Salinity includes all the Cl and Na salts and the $MgSO_4$. Therefore, the potential salinity of an irrigation water can be estimated from the sum of the concentration of the Cl and half of the SO_4 ions:

Potential salinity of irrigation water = $Cl + 1/2\,SO_4$ in me/l.

Classification for potential salinity of irrigation waters with and without restrictive drainage is given in Table 5. Potential salinity is different from "effective salinity" because the former assumes that the Ca from the exchange complex will precipitate as Ca-$MgCO_3$ and $CaSO_4$ even when Ca is deficient in the water (Table 4). These salts are not sufficiently soluble to produce a saline soil condition, whereas all Cl salts and Mg-$NaSO_4$ are highly soluble. The formula Cl + 1/2 SO_4 as an indicator of the salinity of an irrigation water was suggested by EATON (1954) and by additional work of DUTT and DONEEN (1963). They showed that the accumulation of soluble salts from waters containing various cation combinations of SO_4 applied to soils having different ratios of Ca and Mg on the exchange complex caused approximately half of the SO_4 to be precipitated as $CaSO_4$ while the other half remained in the soluble form of Na-$MgSO_4$ in the soil. One exception is when the soil exchange is saturated with Ca which results in nearly all of the SO_4 being precipitated.

Table 5. Classification for potential salinity of irrigation waters[a]

Soil conditions	Class 1 me/l	Class 2 me/l	Class 3 me/l
A Little leaching of the soil can be expected due to low percolation rates	< 3	3– 5	5+
B Some leaching but restricted. Deep percolation or drainage slow	< 5	5–10	10+
C Open soils. Deep percolation of water easily accomplished	< 7	7–15	15+
C_1 Open soils. Medium salt tolerant crops	< 10	20	30+
C_2 Open soils. Salt tolerant crops	< 15	25	35+

[a] The table is based on salt sensitive crops except C_1 and C_2.
For relative tolerance of crop plants to salt see USDA Handbook No. 60, p. 67, 1954 (Ref. [39]).

Plant tolerance to salinity was introduced into the classification (Table 5), as C_1 and C_2 after an intensive investigation of extensive areas in India, where saline waters were used on permeable soils. Some waters were above 35 me/l concentration yet fair production was obtained by growing salt tolerant crops. With sufficient experience and known conditions it may be possible to classify the soil conditions A and B (Table 5), according to the salt tolerance of crop plants.

b) Permeability of a soil is influenced by the sodium content of the irrigation water—a fact recognized for many years. Waters having sodium as the predominant cation tend to disperse, i.e., puddle easily when wet and form hard surface clods when dry, resulting in a reduced rate of infiltration. With an extremely high sodium content in either the water or soil, the irrigation water may remain on the surface for days or weeks, and its disappearance may be due more to evaporation than to percolation into the soil.

Recent studies have indicated that soil permeability, as affected by long-term irrigation will be influenced by the total salt concentration of the water and by the sodium and bicarbonate content. These 3 items are incorporated into a formula termed the "permeability index". This index has been empirically developed from a series of experiments conducted in the laboratory, and a series of lysimeter studies using a large number of irrigation waters varying in ionic relationships and concentration. In addition, it has been tested under field conditions. It has been formulated as follows (DONEEN, 1961):

$$\text{permeability index} = \frac{Na + \sqrt{HCO_3}}{Ca + Mg + Na} \times 100, \text{ Ions in me/l.}$$

Water is classified by this index according to Fig. 3 for soils of medium permeability; however, if the soil characteristics are known, this classification can be refined by using Figs. 4 and 5. Fig. 4 is applicable to soils of naturally low permeability such as deep clay or adobe soils, dense or compact subsoils, and light or medium textured soils of inherently low permeability. Fig. 5 is used for soils of naturally high permeability, such as sand or for sandy loam and heavier soils, where the known infiltration rate is relatively high. With highly permeable soils it may be an advantage to lower the infiltration rate and increase the irrigation

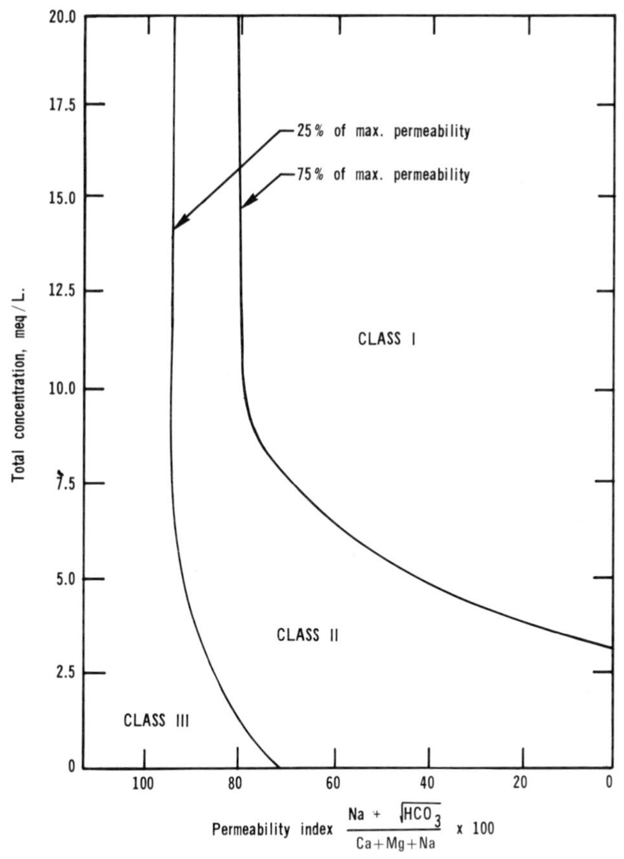

Fig. 3. Classification of irrigation waters for soils of medium permeability. (DONEEN, 1967. Includes other factors such as salt concentration and bicarbonate content of the water)

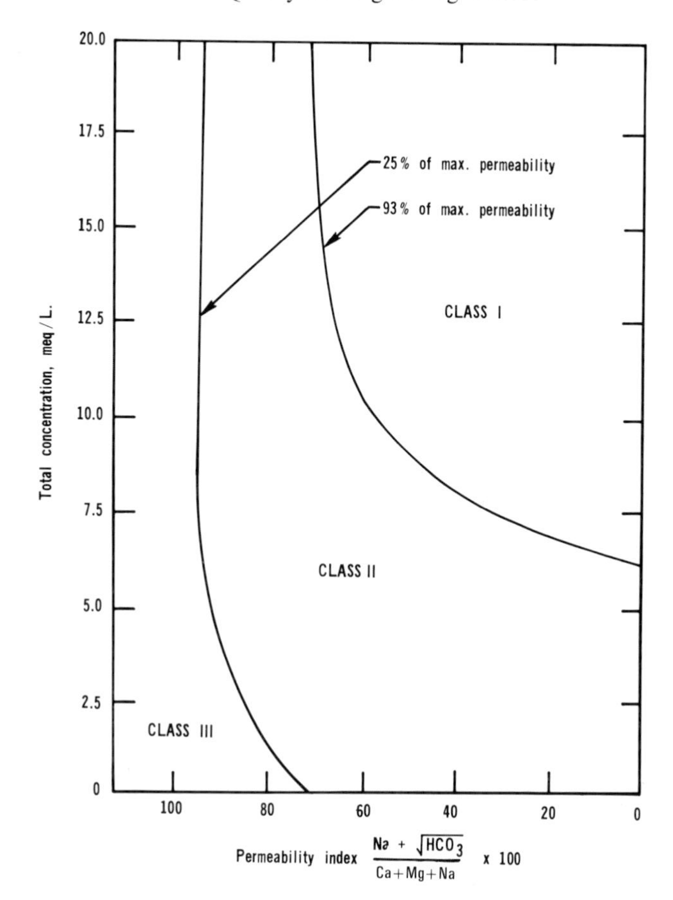

Fig. 4. Classification of irrigation waters for soils of low permeability. (DONEEN, 1967. Includes other factors such as salt concentration and the bicarbonate content of the water)

efficiency by reducing the amount of water percolating to below the rooting depth of the crop.

The maximum permeability listed in Figs. 3, 4 and 5 is based on a concentration of 10–15 me/l of Ca salt, usually $CaCl_2$, according to the results obtained by HENDERSON (1958) and the lysimeter experiments mentioned above.

c) Toxic ions or other materials affecting specific plant species are present in some irrigation waters. Examples are boron, sodium and chloride. They may or may not be related to the accumulation of salts in the soil solution. The principal toxic element is boron. Although it is required for plant growth, the range between beneficial and toxic concentrations, for some plants, is narrow. Boron in toxic concentrations is found in a few localized areas throughout the world. In these areas the ground waters are generally found to be much higher in boron than the surface waters, although some streams or rivers may have toxic concentrations, usually derived from springs or seepage draining into the water course (SCOFIELD and WILCOX, 1931).

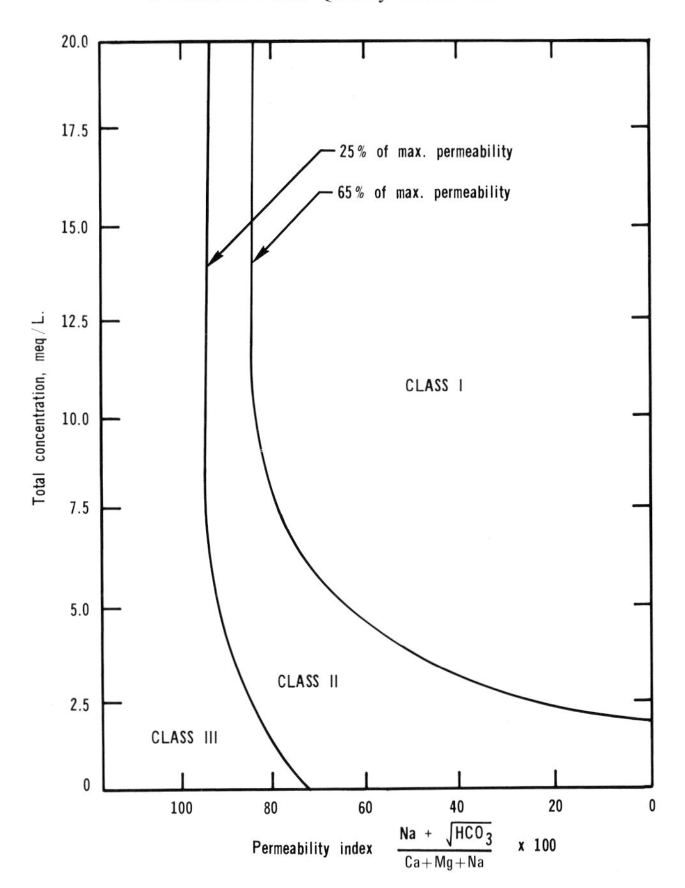

Fig. 5. Classification of irrigation waters for soils of high permeability. (DONEEN, 1967. Includes other factors such as salt concentration and the bicarbonate content of the water)

KELLEY and BROWN (1928) established that boron occurring in natural waters was toxic to citrus and walnuts. In localized non-irrigated low rainfall areas boron may be found in toxic concentrations in the surface layers of soil and their reclamation by leaching is reported by DONEEN (1968). Five classes of boron ranging in concentration from less than 0.25–3.0 ppm and above in the irrigation water were recommended by EATON (1935). Over the next 10 years a series of changes occurred but by consensus evolved the classification most generally used at the present time which is as follows:

	Class		
	1	2	3
Boron ppm	Less than 0.5	0.5–2.0	More than 2.0

These standards are conservative for many situations, as experience has shown that with open soils and good drainage sensitive plants, such as walnuts and citrus, thrive with 1 ppm boron, and at the other extreme many boron tolerant crops do well above 3 ppm.

The toxicity of sodium, chloride, and sulfate has been established for certain plants, but the limits in irrigation water for these ions have not been determined. As discussed above, quality classification for irrigation water according to Doneen's criteria are: potential salinity, permeability index and presence of toxic substances.

Regardless of the classification system for irrigation waters, it is very difficult to establish standards for each class, as there will be much overlapping at the boundaries. Under certain conditions a Class 2 water may be used as successfully as a Class 1. On the other hand, a Class 2 may be as poor as a Class 3 water. At best a classification scheme can be used only as a general guide, and local conditions, to some extent, determine its usefulness. Doneen's recent classification includes a number of factors not included in the earlier ones; namely, the precipitation of slightly soluble salts upon salinization, the role of bicarbonate in the formation of sodic soils and its effect on permeability, the effect of ionic concentration on permeability, the soil characteristics and plant tolerance to salts.

E. Application of Water Quality Criteria to Field Conditions

In applying water quality criteria to field conditions one should take into account the quantity of available water, its management and control, irrigation practices, the crop, and the effect of the irrigation water on drainage and ground water. The following will indicate some of these considerations.

I. Salt Removal by Plants

From time to time it has been suggested that salt tolerant plants will absorb the salt from the soil or water and it can be removed in the harvested crop. That this is a fallacy will be easily recognized if the yield and mineral composition of the crop and the quantity of salt added in the irrigation water are known. LYERLY and LONGENECKER (1957) found for 8 different crops, that an average of 202 pounds of salt present in irrigation water was removed per acre. However, the 202 pounds contained only 22 and 38 pounds of Na and Cl respectively, for a total of 60 pounds of these undesirable ions. Assuming 3 acre-feet of irrigation water was applied containing 10 me/l salt, then about 6000 pounds of salt was added to the soil and approximately 3.4%, or 0.34 me/l removed by the crop; and this does not include the dissolution of salts in the soil profile. Even with a crop that thrives under salty conditions, it is questionable whether even 1 me/l salt would be removed in the harvested portion of the plant.

II. Lime Precipitation in the Soil

With HCO_3 waters, unless frequent heavy leaching of the soil occurs, lime will accumulate in the soil. In addition to the sodium effects, the precipitated $CaCO_3$

will neutralize an acid soil to produce an alkaline soil. In the case of a neutral or slightly alkaline soil HCO_3 waters may greatly increase the lime content, as reported by DONEEN (1959) who showed that this increase in lime may be deleterious to certain crops that grow better on an acid or neutral soil.

III. Foliar Absorption of Salts

It is known that nutrients and other ions can be absorbed through plant leaves. When water containing as little as 5 me/l of Na or Cl is applied by sprinkler irrigation, under high evaporating conditions, sufficient amounts of these ions may be absorbed by the leaves as to cause tip and marginal burn and the plant may eventually be defoliated. Crops known to be sensitive to damage by leaf absorption are some of the stone fruits, almonds, citrus, walnuts and some woody ornamentals. In general, most of the annual crops are less prone and some are quite resistant to foliar injury from sprinkler irrigation.

IV. Amendments to Improve the Quality of Water

Waters of low salt concentration and high Na percentage or having residual Na_2CO_3 can be improved by the addition of Ca salts as demonstrated by DONEEN (1948) and AXTELL and DONEEN (1949). The addition of gypsum to irrigation water greatly improved its infiltration rate. Two factors were involved, (1) the residual Na_2CO_3 was eliminated, and (2) the salt concentration of the water was increased. The average increase in infiltration rate for the 6 waters in this study was 90 percent. Other soluble Ca salts can be used, but gypsum is usually the most economical and readily available. Because gypsum dissolves slowly its addition to water is usually limited to 2–4 me/l. However, with much agitation larger amounts may be dissolved. Machines have been developed for the addition of gypsum in proportion to the flow of irrigation water. The greatest response to gypsum has been with waters in the lower salt content, or less than 10 me/l, whereas with waters of higher concentration gypsum has minimal effects on infiltration because of the high salt content already present in the water and its smaller effect on the composition of the water. Because gypsum dissolves slowly with a limited solubility of about 30 me/l there may be an advantage in applying it directly to the soil to improve the quality of irrigation water. However, this usually requires a larger amount of gypsum than if it is applied directly into the water.

In the acidification of irrigation water for better penetrability, farmers have used 3 products: carbon dioxide (CO_2), sulfuric acid (H_2SO_4), and sulfur dioxide (SO_2). DONEEN (1951) investigated these materials in the field and laboratory. The CO_2 from the exhaust of diesel engines or high pressure cylinders will temporarily decrease the pH of the water near the point of injection, as a small amount of CO_2 will form carbonic acid H_2CO_3, but as the water flows into the furrows or spreads across the land this entrapped CO_2 is quickly dissipated to the atmosphere and the water returns to its normal pH, with no increase in infiltration rates.

Sulfuric acid or any other strong acid will be affected by the buffering capacity of the water according to its HCO_3 content. To neutralize each me/l of HCO_3 in the water, 136 pounds (61.9 kg) of H_2SO_4 is required per acre-foot of water (12.3 cm in depth per hectare). After neutralization of the water, the acidity in the water will increase in proportion to the amount of acid added. In neutralizing the HCO_3 of the water the total concentration does not change but merely shifts from the HCO_3 to the SO_4 form, and with residual Na_2CO_3 in the water this may be beneficial. It is usually not economical to use H_2SO_4 for the treatment of water, and in high concentrations it will be deleterious to concrete pipe and corrosive to irrigation equipment.

Sulfur dioxide may be obtained as a compressed gas in cylinders, from burning sulfur, and from diesel engines burning fuel high in sulfur. The reaction is the same as for H_2SO_4. When the SO_2 gas is entrapped in water it forms sulfurous acid (H_2SO_3), but the molecular weight of SO_2 being less than H_2SO_4, 87 pounds (39.5 kg) of the gas is required to neutralize 1 me/l of HCO_3 in an acre-foot (12.3 cm in depth per hectare) of water. Wetting agents and synthetic detergent products have been applied to irrigation water to improve the infiltration rate but so far these materials have not been successful.

V. Salt Balance and Leaching Requirements

In order to maintain a salt balance a quantity of dissolved salts must be removed in the drainage water which is equal to that carried into the area with the irrigation water. An unfavorable salt balance occurs when more salts are brought into the area than are removed by drainage, thus producing a saline soil condition. The original concept of salt balance was for large areas, or irrigation projects, but recently the concept has been applied to any size unit including individual farms or small tracts of land where the salt balance pertains to the depth of soil penetrated by plant roots either in the presence or absence of a high water table.

Leaching of the soil is required to maintain a salt balance. The leaching requirement has been defined by the U.S. Salinity Laboratory Staff (1954) as the fraction (or percentage) of irrigation water that must be leached through the root zone to keep the salinity of the soil below a specific value. Recently RHOADES (1972) has reviewed the U.S. Salinity Laboratory's original concept on water quality and leaching requirement and has suggested improvements and areas requiring additional work.

The purpose of this discussion is not to delve into the concepts of the leaching requirement but to point out some of the water quality factors having a bearing on this concept. The leaching requirement is based on the total salt concentration of the irrigation water and that being leached below the rooting zone, thus ignoring the precipitation of the less soluble salts. Our recent studies show that using the formula $Cl + 1/2\,SO_4$ in me/l, gives better results than the use of total salts in calculating the leaching requirement. In addition, the leaching requirement assumes a linear relationship between the concentration of the irrigation water and of the solution leaving the rooting zone. This is not necessarily true under field conditions, where the surface soil is usually in equilibrium with the irrigation water and accumulation of salts occurs in the subsoil, as shown by DONEEN (1960) in lysimeter and field studies. On the west side of the San Joaquin Valley, California, where waters are of medium quality and are expensive, in order to increase irrigation efficiency, the farmer's general practice is to wet the soil about a meter or less in depth during the irrigation season and between crops to apply sufficient water to leach the accumulated salines into the subsoil, below the depth of plant rooting.

The leaching requirement is automatically provided in permeable soils when water percolates below the rooting depth when the crop is irrigated—i.e., under conditions of low irrigation efficiency. Leaching is usually not required where the seasonal rainfall wets the soil to an average of 2 meters or more, unless the water has a high concentration of salts. This is the case under Mediterranean or monsoon climates.

VI. Ground Water Recharge

In irrigated areas where the water supply is pumped from a ground water basin, the quality of the replenishment water will be influenced by the concentration of the salts leaving the rooting zone, the rainfall, and the contribution from the non-irrigated areas of the basin. In low rainfall areas the soluble salts native to the soil profile between the soil surface and the water table will cause a deterioration of the ground water as shown by DUTT and TANJI (1962) and DONEEN et al. (1967). In many ground water basins the height of the water table may drop or it may be depleted. It then becomes necessary to replenish the ground water by artificial means, such as by using excess amounts of surface water, when available, for irrigation, by setting up percolating ponds or recharge areas, or by cyclic ground water storage operations.

For recharge, the concentration of the percolating water may be calculated as the total of $Cl + SO_4 + 2$ me/l HCO_3. Precipitation of gypsum will not occur in irrigation waters having an appreciable amount of sulfate until the soil solution has reached a concentration of more than 3 millimhos, or 30 me/l of gypsum. In irrigation practice where salts accumulate due to the reasons listed above, most salts will be periodically removed by leaching, occurring between crops in the pre-irrigation, before annuals are planted, or during the winter season for the perennials. With this leaching, the precipitated gypsum according to DUTT (1964) will either dissolve and enter the drainage system, or percolate to the ground water.

Since the precipitated carbonate is less soluble, only a small amount will be dissolved in the leaching water. Most saline soil solutions, or irrigation water having an appreciable amount of salt, usually contain less than 15% bicarbonate ions—in the range of 1–4 me/l. Therefore, the assumption can be made that about 2 me/l of bicarbonate will appear in the drainage or deep percolating waters. However, waters having a low salt concentration, which percolate through non-saline lime soil, may have their bicarbonate content increased to a level of 4–9 me/l.

VII. Sewage and Industrial Wastes

Sewage and industrial wastes are often used in the production of crops. These waste waters may be discharged into rivers or water courses, used for irrigation of crops, or ponded and allowed to percolate to the ground water. Their quality for irrigation should be judged by the amount and type of salts they contain. The organic matter present will be readily oxidized in the surface soil. According to the research available, the detergent, soft or hard, known as alkyl benzene sulfonate (ABS) is not detrimental to plant growth (Irrigation of Agriculture Lands, 1967). The chlorinated hydrocarbons which may occur in some waste waters are mostly fixed in the soil and usually reach the ground water only in minute quantities. The handling of waste water is a specialized field, involving treatment plants, water management, etc., with many related studies which are beyond the scope of this chapter.

F. Conclusions

This chapter deals primarily with the arid regions of the world, where irrigation is required for economical production of crops. However, the quality requirements should be applicable to some of the high rainfall areas, where supplemental irrigation is practiced.

In general the better quality waters are those having a low concentration of salts. However, natural surface and ground waters usually contain salts in varying degrees of concentration, and water quality requirements are based on the principle of salt accumulation and its effect on soil structure or permeability, and on plant growth.

The farmer rarely has a choice of different qualities of irrigation water. In most cases he has only one source, whether it is good or extremely poor. Consequently, water should seldom be condemned for use in irrigation as the farmer's livelihood usually depends on it; besides we have been wrong too many times in the past. Instead, indicate to the farmer that the quality of the irrigation water is poor; suggest amendments if needed, crops that can be grown most successfully, and water management practices that should be followed. When these factors are combined into a working program, many very poor quality waters can be used successfully in agriculture.

References

AXTELL, J. D., DONEEN, L. D.: The use of gypsum in irrigation water. Better Crops with Plant Food **33**, No. 9, 16–23 (1949).

California State Water Quality Control Board.: Water-Quality Criteria Publ. No. **3**, 1–548 (1952).

DONEEN, L. D.: The quality of irrigation water and soil permeability. Proc. Am. Soil Sci. Soc. **13**, 523–526 (1948).

DONEEN, L. D.: Unpublished data, Department of Water Science, and Engineering. Davis: University of California 1951.

DONEEN, L. D.: Salination of soil by salts in the irrigation water. Trans. Am. Geophys. Union **35**, 60, 943–950 (1954).
DONEEN, L. D.: Studies on water quality criteria. Univ. of Calif. Water Resources Center. Contribution **14**, 46–56 (1958).
DONEEN, L. D.: Evaluating the quality of irrigation water in Ventura County. Calif. State Dept. of Water Resources Bull. **75**, V. 2, F1–F33 (1959).
DONEEN, L. D.: Water quality in relation to soil and crop production, lower San Joaquin Valley. Calif. State Dept. of Water Resources Bull. **89**, C1–C125 (1960).
DONEEN, L. D.: In: SCHIFF, L. (Ed.): The influence of crop and soil on percolating waters. Proc. Biennial Conference on Ground Water Recharge (1961).
DONEEN, L. D.: Water quality requirements for agriculture. The National Symposium on Quality Standards for National Waters. Education Series No. 161, Univ. Mich. Ann Arbor 213–218 (1967).
DONEEN, L. D. (Ed.): Agricultural development of new lands west side of San Joaquin Valley. Dean's committee Report **1**, 1–83; **2**, 1–98 (1968).
DONEEN, L. D.: Report on salinity, alkalinity, and water management. Tamil Nadu Agric. Univ. Coimbatore, India. General Report **1**, 13–25 (1970).
DONEEN, L. D., HENDERSON, D. W.: Quality of irrigation water and chemical and physical properties of soil. Trans. 7th Intern. Congress of Soil Sci. **1**, 516–522 (1960).
DONEEN, L. D., TANJI, K. K., DUTT, G. R., PAUL, J. L.: Quality of percolating waters. Hilgardia **38**, 285–353 (1967).
DUTT, G. R.: Effect of small amounts of gypsum in soils on the solutes in effluents. Proc. Am. Soc. Soil Sci. **28**, 754–757 (1964).
DUTT, G. R., DONEEN, L. D.: Predicting the solute composition of the saturation extract from soil undergoing salinization. Proc. Am. Soil Sci. Soc. **27**, 627–630 (1963).
DUTT, G. R., TANJI, K. K.: Predicting concentrations of solute in water percolated through a column of soil. J. Geophys. Res. **67**, 3437–3439 (1962).
EATON, F. M.: Boron in soils and irrigation waters and its effect on plants. U.S. Dept. Agr. Tech. Bul. **448**, 1–132 (1935).
EATON, F. M.: Irrigation agriculture along the Nile and the Euphrates. The Sci. Monthly **69**, 34–42 (1949).
EATON, F. M.: Significance of carbonates in irrigation waters. Soil Sci. **60**, 123–133 (1949a).
EATON, F. M.: Formulas for estimating leaching and gypsum requirements of irrigation waters. Texas Agric. Expt. Sta. Misc. Pub. **111**, 1–18 (1954).
FIREMAN, M.: Permeability measurements on disturbed soil samples. Soil Sci. **58**, 337–353 (1944).
FIREMAN, M., BODMAN, G. B.: The effect of saline irrigation upon the permeability and base status of soils. Proc. Soil. Sci. Soc. Am. **4**, 71–77 (1939).
HEM, J. D.: Study and interpretation of the chemical characteristics of natural water. U.S. Geog. Survey Water Supply Paper **1473**, 1–269 (1959).
HENDERSON, D. W.: Influence on soil permeability of total concentration and sodium in irrigation water. Univ. of California Water Resources Center, Contribution **14**, 153–157 (1958).
HILGARD, E. W.: Soils, their formation, properties, composition, and relations to climate and plant growth. New York: MacMillan 1906.
International Source Book on Irrigation and Drainage of arid lands, FAO/UNESCO, 264–281 (1973).
Irrigation of Agricultural Lands, Am. Soc. Agron., Madison, Wisc., Monograph **11**, 104–122 (1967).
KELLEY, W. A., BROWN, S. M.: Boron in the soils and irrigation water of Southern California and its relation to citrus and walnut culture. Hilgardia **3**, 445–458 (1928).
LUNT, O. R.: Proceedings of agricultural water quality research conference. Univ. Calif. Water Resource Center Rept. **5**, 1–69 (1963).
LYERLY, P. J., LONGENECKER, D. E.: Salinity control in irrigated agriculture. Texas Agr. Expt. Bul. **876**, 1–19 (1957).
Physical Edaphology, the physics of irrigated and non-irrigated soils, pp. 451–472. San Francisco: W. H. Freeman and Co. 1972.

QUIRK, J. D., SCHOFIELD, R. K.: The effect of electrolyte concentrations on soil permeability. Soil Sci. **6**, 163–178 (1955).
RHOADES, J. D.: Quality of water for irrigation. Soil. Sci. **113**, 277–284 (1972).
SCOFIELD, C. S.: The salinity of irrigation water. Smithsn. Inst. Ann. Rept. **1935**, 275–287 (1936).
SCOFIELD, C. S., WILCOX, L. V.: Boron in irrigation waters. U.S. Dept. Agr. Tech. Bull. **264**, 1–65 (1931).
STABLER, H.: Some stream waters of the Western United States. U.S. Geol. Survey Water-Supply Paper **274**, 188 (1911).
TANJI, K. K., DONEEN, L. D.: A computer technique for prediction of $CaCO_3$ precipitation in HCO_3 salt solutions. Proc. Soil Sci. Soc. Am. **30**, 53–56 (1966).
U.S. Salinity Laboratory Staff.: Diagnosis and improvement of saline and alkali soils. U.S. Dept. Agr. Handbook **60**, 69–82 (1954).
WILCOX, L. V.: The quality of water for irrigation. U.S. Dept. Agric. Tech. Bull. **962**, 1–40 (1948).
WILCOX, L. V., BLAIR, G. Y., BOWER, C. A.: Effect of bicarbonates on suitability of water for irrigation. Soil. Sci. **77**, 259–266 (1954).

Chapter V

Effects of Land Use on Salt Distribution in the Soil

A. J. Peck

A. Introduction

Saline soils are known to exist in many places throughout the world, particularly in arid and semi-arid zones. The great majority of these areas appear to have been affected by excess soluble salts for thousands of years, but there is a very large area of land which has become saline within the period of recorded history. The evidence suggests that these areas of secondary salinity result from man's activities, or in the language of this decade, secondary salinity is an environmental problem.

Indeed it can be argued that secondary salinity was the first great environmental problem faced by man. Jacobsen and Adams (1958) consider that beginning as early as 2400 BC, increasing soil salinity played an important part in the breakup of the ancient Sumerian civilization.

Secondary salinity appears to be most common in irrigated areas, affecting as much as 4×10^7 ha, or one-third of the world's irrigated land (Reeve and Fireman, 1967). This has led to some doubt about the permanence of economically feasible irrigation agriculture in the Western USA and elsewhere (Reeve and Fireman, 1967).

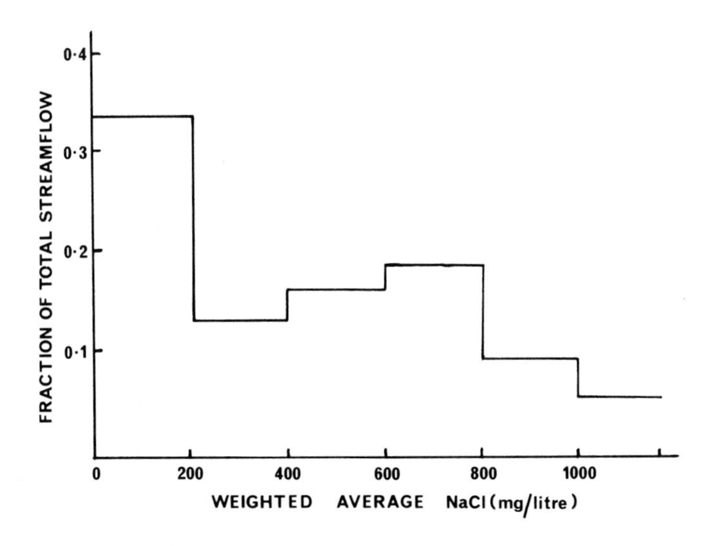

Fig. 1. The fraction of usable streamflow in south-western Australia at present in various salinity classes. Data from records of the Public Works Department of Western Australia

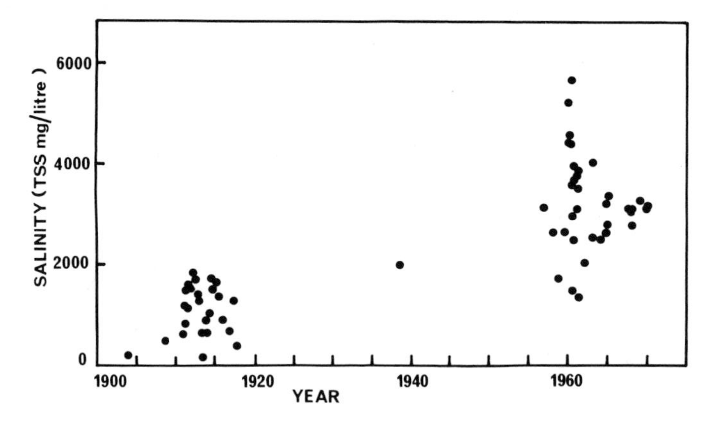

Fig. 2. The trend in salinity of the Blackwood River, largest stream in south-western Australia, at Bridgetown. Data from WOOD (1924), and the Western Australian Department of Agriculture

Not insignificant areas of secondary salinity occur in association with dryland agricultural practices too. In the short history of European settlement in southern Australia, an estimated 2×10^5 ha of land which first supported good crop or pasture is now suitable only for halophytic species (NORTHCOTE and SKENE, 1972). The problem also occurs in Canada (GREENLEE et al., 1968), and in the northern USA where the affected area of 3.2×10^4 ha is estimated to be increasing by 10% each year (CLARK, 1971).

In most cases where soils are affected by secondary salinity there is also an increased salinity of water in the local drainage system whether this be an artificial system or a natural stream. In an area where good quality water supplies are sparse, an increase of stream salinity may be of greater economic and environmental consequence than the reduced agricultural productivity of saline soil. For example, in south-western Australia the area of secondary salinity, about 1.6×10^5 ha, is less than 1% of the total area of farmed land (LIGHTFOOT et al., 1964). But in this same area about 50% of all streamflow fails to meet the conventional standard for drinking water (TDS < 500 mg/l), and there is good evidence of a three-fold increase in salinity of the largest river over the last 50 years. See Fig. 1 and 2. Other data (PECK and HURLE, 1973) and historic records strongly suggest that increases of stream salinity have occurred throughout this region since European settlement began about 100 years ago.

This article examines factors influencing the distribution of soluble salts in the natural environment, and the redistribution of salts which results from environmental disturbances.

B. Sources of Salts

The soluble salts in soils and streams which contribute significantly to total dissolved solids (TDS) and salinity problems consist mainly of the ions Na^+, Ca^{2+}, Mg^{2+}, K^+, Cl^-, SO_4^{2-}, HCO_3^-, and CO_3^{2-}. The relative contributions of the various cations and anions vary considerably from place to place, depending on the importance of the several sources discussed below.

I. Surface Deposition

Soluble salts are found in rainwater, irrigation water, and dust settling on the soil. Other depositional sources are artificial fertilizers, herbicides, insecticides, fungicides, solid wastes, highway de-icing compounds and materials from vehicle exhaust gases. Gaseous exchange with the atmosphere is a primary source for HCO_3 and CO_3.

PECK and HURLE (1973) report estimates of Cl deposition in rain which exceed 100 kg/ha/yr. While this is exceptionally large, ERIKSSON (1960) estimates that atmospheric deposition is an important source of Cl in many parts of the world, particularly those with a maritime climate.

SCOFIELD (1940) is credited with the salt balance concept relating the input of solutes in irrigation water to an area with the loss in drainage water. In the Murrumbidgee Irrigation Areas of south-eastern Australia, which are supplied with exceptionally good water (TDS about 70 mg/l), the mean annual deposition of salts has been estimated as about 200 kg/ha (EVANS, private communication).

In a near coastal region BALDWIN (1971) estimated a deposition of Cl in dry dust of about 30 kg/ha/yr. The net contribution of dust in more arid regions may vary widely. BONYTHON (1956) suggests that erosion of dry salt may be a significant factor in the loss of soluble materials from some areas, while elsewhere deposition has created lunettes very high in $CaSO_4$ in the lee of playa lakes (BETTENAY, 1962).

II. Weathering Reactions

Chemical weathering of rock minerals is usually considered to be the ultimate source of all soluble salts. The reactions taking place may be complex. CLEAVES et al. (1970) discuss their data on the geochemical balance of an experimental watershed, and an interpretation of the weathering reactions taking place. They show that the relative importance of deposition and weathering in a given area may be quite different for different ions. In their watershed HCO_3 was predominantly a weathering product, while Cl and SO_4 were deposited in precipitation.

Weathering reactions may also release salts which do not form part of the rock, but are securely locked in interstices of the rock mass. HOLMES (1971) has discussed an example.

Changes of temperature or pH may lead to the release of solutes from minerals and adsorption sites, or conversely the solutes may be precipitated within, or adsorbed on the particle surfaces of a soil.

III. Redistribution

In many problems of secondary salinity, it appears that there has been little if any disturbance to either rates of deposition of solutes on the soil surface, or the rate of release of solutes by weathering reactions. The accumulation of salts at a particular location is merely a redistribution of solutes within the soil-water system. The operative transport processes are described in the following section, and later sections deal with distributions of solute concentration under various circumstances.

C. Transport of Solutes in Porous Materials

Mechanisms of solute transport in soils and aquifers have been reviewed in detail by GARDNER (1965), PHILIP (1970), and FRIED and COMBARNOUS (1971). Two important situations are usually examined: that when the rate of bulk solution flow is negligible, and that when the solution velocity is significant.

I. Transport without Solution Flow

In this case solute transport takes place by molecular diffussion in response to a gradient in concentration of the particular ion in the soil solution. The effective diffusion coefficient, however, differs from that in bulk solution due to several effects such as the reduced cross section for diffusion and increased tortuosity of the path in soil. GARDNER (1965) relates diffusion coefficients in soils to those in bulk solution, and lists values of the coefficient for Cl diffusion in several soils. In transient situations other processes such as dead-end pores and exchange reactions with other ions on soil particle surfaces also contribute, and may further reduce the effective diffusion coefficient in soil below that in bulk solution.

Although diffusion may be important in some particular situations, the magnitude of this transport coefficient is such that flux densities of salts are usually very small when there is no significant movement of the bulk solution.

II. Interactions of Diffusion and Convection

When water moves through a soil or aquifer material the complex geometry of the fluid-filled pore space results in a spectrum of microscopic solution velocities. This results in the dispersion of a solute initially concentrated in a small part of the medium. Molecular diffusion also operates, and over a range of solution velocities there is a significant interaction which contributes to solute dispersion both in the direction of flow, and transverse to it.

FRIED and COMBARNOUS (1971) show that an important parameter of this dispersion process is the Peclet number Pe defined by

$$Pe = |u| l / D, \tag{1}$$

where $|u|$ is the magnitude of the average solution velocity, l is a characteristic pore dimension, and D is the solute diffusion coefficient in bulk solution. As the Peclet number increases above about 1, the coefficients for both the longitudinal and transverse components of dispersion become larger than the coefficient of molecular diffusion in the medium.

Often the Peclet number is less than 1 in soil-water flows so that molecular diffusion remains the dominant mechanism for mixing at an interface of solution concentration. However the interface will be transported with the mean solution flow which is usually the dominant mechanism for solute transport over appreciable distance.

Groundwater movement in an aquifer which yields water rapidly is often characterized by a Peclet number much greater than 1 so that solute mixing at an interface is more rapid than it would be by molecular diffusion alone.

Other processes affect solute transport in many soils. BIGGAR and NIELSEN (1967) showed that exchange reactions retarded the passage of Mg in a Ca-saturated sandy soil, while PASSIOURA and ROSE (1971) showed that significant structure in the medium will enhance the role of convection in solute dispersion.

D. Equilibrium Distribution of Solute Concentration

In an equilibrium situation the inputs of salts by deposition and weathering in an area will be balanced by exports of the solutes in streamflow, underground water losses, and wind erosion. For example, the data of Table 1 suggest something close to a Cl balance in some forested watersheds. But at any instant appreciable quantities of soluble solids may be distributed within an area. Table 2 lists total quantities of soluble solids in the soil-water system averaged over several locations in south-western Australia, including some of the watersheds of Table 1.

The spatial distribution of salts in some parts of the environment can be related to patterns of water movement, evaporation and transpiration.

Table 1. Input (in rainfall) and loss (in streamflow) of chloride for some forested catchments in south-western Australia. For details of methods used to obtain these data, see PECK and HURLE (1973)

Catchment name	Chloride input kg/ha/yr	Chloride loss kg/ha/yr
Julimar	53	78
Seldom Seen	120	160
More Seldom Seen	120	140
Waterfall Gully	110	180
North Dandalup	130	180
Davies	130	140
Yarragil	97	130
Harris	84	130

Table 2. Average storage of soluble solids in the soil profile at sites in south-western Australia. Data from G. M. DIMMOCK (private communication)

Rainfall zone mm/yr	Number of sites sampled	Average depth to basement m	Average salt storage kg/ha
< 800	29	21	8.1×10^5
$800 \rightarrow 1000$	4	22	2.4×10^5
> 1000	10	21	1.7×10^5

I. Vertical Distribution in the Root Zone

We have noted that rain or irrigation water entering the soil surface carries soluble salts. These are concentrated within the soil-water system as a result of water uptake by plant roots. GARDNER (1967) has examined the equilibrium salt profile when water of salinity c_o is supplied to the soil at rate P, and removed by roots at the local rate w. In the simplest situation there is no other source of salt, there is complete exclusion of salt during water uptake by roots, and soil water always moves downwards to a deep water table. Then the equilibrium salt profile is given by

$$c = Pc_0 / \left(P - \int_0^z w dz \right), \qquad (2)$$

where c is concentration of salts in the soil solution at depth z measured downwards from the soil surface.

So long as w is positive, Eq. (2) describes a curve of salinity increasing with depth. However, this does not always occur. For example, Fig. 3 shows the Cl profile at a forested site in south-western Australia where it is presently believed that water moves vertically downwards, and near steady-state conditions prevail. The maximum of salinity well above the water table is a feature of this, and other profiles in the region, which cannot be predicted by the present analysis. This suggests that further development of the analysis is required.

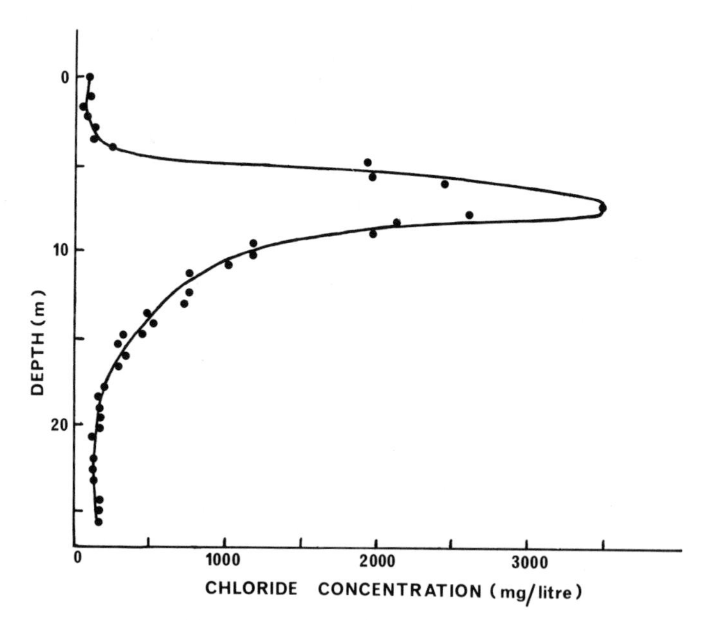

Fig. 3. Concentration of chloride in the soil solution as a function of depth at a forested site in south-western Australia. Note the peak in concentration well below the surface and above the water table, which is common in this area

II. Vertical Distribution above a Shallow Water Table

Water moves upwards by capillarity from a shallow water table to the soil surface where evaporation concentrates soluble salts and may lead to their deposition. The steady upward movement from a fixed water table has been studied in considerable detail (HILLEL, 1971). As the depth of the water table (measured from the soil surface) increases, the steady rate of evaporation that can be sustained decreases quite rapidly. With very few exceptions the rate of evaporation that can be sustained from a water table 2 m or more below the surface is less than 1 mm/day (TALSMA, 1963). HADAS and HILLEL (1972) found that a non-uniform soil profile further reduces the rate of upward flow, particularly when the surface soil is relatively coarse.

The enhanced solute concentration at the soil surface results in a diffusive transport of salt downwards in opposition to the upward convective movement. In the steady state, GARDNER (1965) shows that the resultant salt profile beneath a bare soil is

$$c = c_0 \exp(-Ez/D), \tag{3}$$

where c is concentration in the soil solution at depth z (measured from the surface), c_0 is that at the surface, E is the steady rate of evaporation, and D is the diffusion or dispersion coefficient. The exponential decay of concentration with depth is supported by experimental data presented by GARDNER.

III. Areal Distribution

The distribution of groundwater salinity within a watershed area reflects the distribution of sources and sinks of water and the solutes.

Some generalizations are possible on patterns of water movement in small drainage basins following studies by TOTH (1963), FREEZE and WITHERSPOON (1967, 1968), and others. Often the water table or piezometric surface is topographically similar to the soil surface. It follows that uplands tend to be areas of recharge of groundwater which tends to be discharged by evapotranspiration or liquid seepage into springs and streams in the valleys and lowlands. The movement of water from recharge to discharge sites transports solutes, and a salinity distribution develops according to the location and strength of sources and sinks of the salts.

A relatively simple situation exists when the major source of salts is deposition at a rate which is a known function of location. ERIKSSON and KHUNAKASEM (1969) assumed an equilibrium distribution of Cl in the Coastal Plain aquifer in Israel, and calculated the distribution of intensity of groundwater recharge from known distributions of Cl deposition rate and concentration in the groundwater. If the recharge rate was a known function of location, this method could be inverted to estimate Cl concentration in groundwater across the Plain.

More recent and detailed investigations of solute concentration and changes of ionic composition in groundwater on a regional scale have been reported by SCHWARTZ and DOMENICO (1973).

E. Factors Which May Disturb an Equilibrium Distribution

In general, changes in the distribution or strength of sources and sinks of water and solutes will disturb an equilibrium distribution of salinity. Man's influence on sources of salts seems to be relatively small in many cases, and the most widespread secondary salinity problems are associated with changes in the pattern and rates of water movement; that is, with changes in the local hydrologic regime. Some exceptions are salinity problems associated with disposal of wastes, but these are not considered further here.

The practice of irrigation is an obvious disturbance of the local water balance. In naturally arid regions, the amount of irrigation water applied may greatly exceed rainfall. The Murrumbidgee Irrigation Areas in south-eastern Australia are semi-arid with yearly rainfall of about 420 mm. This is supplemented by an average 360 mm of irrigation water, so the total input of water to the soil has been nearly doubled by the irrigation development.

Perhaps the most widespread change of the local water balance results from removal of natural vegetation, and the introduction of agricultural plants and management practices, or of exotic forest species. For example, Table 3 lists estimated groundwater recharge for areas in south-western Australia which were vegetated by evergreen forest (predominantly *Eucalyptus* species), and now carry annual crops and pastures. Similarly, greater recharge under grass than forest has been reported from investigations in many other parts of the world (BOUGHTON, 1970; URIE, 1971).

In grain producing areas of the United States and Canada the development of secondary salinity is attributed to the use of fallow as a management practice to conserve soil moisture (CLARK, 1971). It has been shown too that in those areas crop water use, and by implication local groundwater recharge, is significantly affected by fertilizer application. For example, BROWN (1971) reports that total water use by a wheat crop increased by 94 mm in response to the application of fertilizer nitrogen at the rate 268 kg/ha.

A more complex and unpredictable disturbance to the local water balance can be expected from strip-mining operations, such as bauxite mining, which completely remove a hydrologically active part of the soil profile. In this case, not only

Table 3. Estimated groundwater recharge before and after forest removal for farm operations in catchments in south-western Australia. For details of methods used to obtain these data, see PECK and HURLE (1973)

Catchment name	Recharge before clearing mm/yr	Present recharge of farmland mm/yr
Brockman	8	73
Wooroloo	4	61
Dale	0.8	24
Hotham	2	26
Williams	1	37
Collie East	2	60
Brunswick	70	500

is material exposed which is likely to have quite different infiltration properties than the original soil, but local surface drainage patterns are usually modified too.

Although disturbance of the local water balance has been emphasized above, a salinity problem could result from a change of the vegetation cover which had no effect on the water balance. Equation (2) relates the salt profile to the pattern of water uptake by plant roots $w(z)$. It follows that any change of $w(z)$, such as the introduction of plants with a different rooting behavior, will affect salt storage in the profile even though total plant water use $\left(\int_0^\infty w dz\right)$ may be unchanged. If salt storage in the profile decreases, then the solutes released may present a transient salinity problem elsewhere.

F. Dynamic Response to Increased Recharge

With few exceptions the development of secondary salinity is a result of increased groundwater recharge following either application of irrigation water or a vegetation change which reduces plant water use. The situation of interest then, is the behavior of the soil-water system, including solute concentration, in response to increased recharge over part or all of an area in which there is initially an equilibrium distribution of soluble salts.

There is some advantage in separating the discussion of changes in the soil-water system from that of the associated transport and redistribution of solutes.

I. Water Table Changes

Some delay must be expected between an increased recharge at the soil surface, and any response of the water table. For simplicity, let us assume that the recharge rate is initially a steady value R_0, and increases abruptly to another steady value R_1. Infiltration theory (PHILIP, 1969) indicates that the rate of penetration of the resultant wetting front will approach a constant value u given by

$$u=(R_1-R_0)/(\theta_1-\theta_0), \tag{4}$$

where θ_1 is the volumetric moisture content at hydraulic conductivity $K_1=R_1$, and θ_0 is that at $K_0=R_0$. Values of u calculated for several soils and changes of the recharge rate are listed in Table 4. The lower step of recharge (5–50 mm/yr) could result from a changed land use from forest to pasture (see Table 3), and the larger change (50–500 mm/yr) could follow an irrigation development. Note that the rate of penetration of the wetting front is not simply proportional to the increase of recharge rate. The non-linearity arises through the relationship between hydraulic conductivity K and volumetric moisture content θ.

When the increased recharge reaches the water table, its response depends on aquifer characteristics and distance to a discharge area such as a spring or gaining stream. A relevant problem has been discussed by TSENG and RAGAN (1973). They show that the characteristic time for water table response is T_{wt} given by

$$T_{wt}=(\theta_s-\theta_1)L^2/K_s h, \tag{5}$$

Table 4. Calculated values of the rate of penetration of the wetting front, u, and the maximum rate at which the water table may rise, $-dZ/dt$, following increased recharge (R_0 to R_1) for several soils. Characteristics of the soils were obtained from the references listed within this table. Methods of calculation are described in the text

Soil	R_0 mm/yr		R_1 mm/yr	u m/yr	$-dZ/dt$ m/yr
Ida silt loam (KUNZE, UEHARA and GRAHAM, 1968)	5	→	50	0.9	0.2
	50	→	500	5.6	3.1
Chino clay (GARDNER, 1960)	5	→	50	0.8	0.3
	50	→	500	3.8	6.7
Indio loam (GARDNER, 1960)	5	→	50	0.3	0.2
	50	→	500	3.8	3.0
Pachappa sandy loam (GARDNER, 1960)	5	→	50	1.0	0.2
	50	→	500	7.5	2.1

where K_s is the hydraulic conductivity of the water-saturated soil (volumetric moisture content θ_s), L is distance from the water table divide to the line of discharge, and h is the depth of saturation of the soil over an effectively impervious substratum.

In the limiting situation ground water movement is negligible, and then the water table would rise at a steady rate—dZ/dt given by

$$-dZ/dt = R_1/(\theta_s - \theta_1), \tag{6}$$

where Z is depth to the water table from the soil surface. This limiting rate should be a reasonable approximation to the actual rate when the water table just begins to rise, and T_{wt} is several years or more. Values of $-dZ/dt$ are also listed in Table 4.

In some cases the increase of recharge may be temporary, returning after a relatively short time to its pristine value. For instance, this would be expected when an area of forest was clear-felled for lumber, and then replanted with small trees which, at least for some years, had smaller water requirements than the original forest. A temporary and local increase of recharge will give rise to a groundwater mound. Development and dissipation of groundwater mounds has been investigated by BITTINGER and TRELEASE (1965). With some simplifications the characteristic time of decay of the mound is given by Eq. (5), but in this case L is the radius for a circular mound.

Should the area of recharge supply a confined aquifer system, the term $(\theta_s - \theta_1)$ in Eq. (5) is replaced by the storage coefficient (change of volume of storage per unit area per unit change of fluid pressure). This is numerically a much smaller term, and the time scale for adjustment of the groundwater system will be proportionally reduced.

As an example, consider an irrigation development ($R_0 = 50$ mm/yr; $R_1 = 500$ mm/yr) on Ida silt loam ($K_s = 10^2$ m/yr) above the groundwater divide which is 500 m from a stream ($L = 500$ m). Let the water table be initially at a depth of 10 m, and let there be 25 m of the saturated soil overlying an impermeable stratum ($h = 25$ m).

From Table 4, the time taken for first response of the water table will be about $10/5.6 \approx 2$ years. The water table will begin to rise at a rate of about 3 m/yr so that the least time for elevation of the water table to within 2 m of the soil surface is $8/3 \approx 3$ years, or 5 years after irrigation began. The characteristic time for water table response is [by Eq. (5) and $\theta_s - \theta_1 = 0.16$] $T_{wt} = 16$ yrs, so that a period of 50–100 years would pass before the water table became steady at a new level, and discharge to the stream reached a new equilibrium.

It can be shown that in this example the water table would rise to the soil surface if recharge was uniform from the divide to the stream. Installation of a suitable artificial drainage system could control the water table at a depth of say 2 m, but the local hydrologic regime would still be disturbed by the 8 m rise of water table from its pristine position. Consequently discharge of groundwater to the stream would still increase from its pristine value to a new equilibrium.

We may note that increased recharge very close to the stream would have a rapid effect on groundwater discharge to the stream. However, when the recharge site is remote from the stream line, a significant increase of groundwater discharge would lag the increased recharge by a time of order T_{wt}.

II. Groundwater Displacement

When there is an increased recharge into a soil-water system containing saline water, then it may be the latter which is first released in the increased discharge to streams and springs. This is the explanation of saline seepages which have developed in North America and Australia. Table 5 presents Cl inputs (in rainfall) and losses (in streamflow) for several watersheds in south-western Australia which have been cleared of forest vegetation, for farming, within the last 100 years. Similar data for areas which are still forested were presented in Table 1, and estimated changes of recharge in the areas now farmed were listed in Table 3.

The leaching process which accompanies increased recharge has been mentioned in the earlier section on solute transport. On the field scale, details of geometry of the water-bearing material may affect the concentration of solute in

Table 5. Input (in rainfall) and loss (in streamflow) of chloride for some catchments in south-western Australia which include substantial areas of farmland. Generally native forest has been replaced by agricultural plants within the last 100 years. This data may be compared with Table 1 for behavior of catchments which are still forested. For details of methods used to obtain these data, see PECK and HURLE (1973)

Catchment name	Chloride input kg/ha/yr	Chloride loss kg/ha/yr
Brockman	80	340
Wooroloo	78	420
Dale	24	460
Hotham	48	370
Williams	31	650
Collie East	50	740
Brunswick	110	350

the groundwater discharging to a stream. The best generalization available at present comes from experiments by MULQUEEN and KIRKHAM (1972). They used a physical model to investigate the concentration of salt in drainage water when salt was initially distributed uniformly over the soil surface, and there was steady uniform recharge over the surface, and discharge from a series of tile drains. They observed a very rapid build up, and then an exponential decay of solute concentration in the drainage water.

PECK (1973) has shown that the behavior observed in Mulqueen and Kirkham's experiments is in accord with a very simple analysis of the phenomenon. In particular, the time constant of the exponential decay of solute concentration in the drains is T_l given by

$$T_l = V/R, \tag{7}$$

where V is the volume of water stored in the soil-water system below unit area of soil, and as before R is the rate of recharge per unit area. Peck notes that whereas Eq. (7) may be useful to define the characteristic time, there is no reason to believe that there will always be an exponential decay of concentration in the discharging solution. Earlier ERIKSSON (1960) had defined a time constant for equilibration of groundwater salinity in a similar way. In hydrology T_l is often referred to as the conventional residence time.

The method used to estimate T_l implies that solutes move quite freely through the soil-water system. If, however, processes such as the exchange reactions mentioned earlier retard ion transport, then the characteristic time of decay of concentration of that ion in drainage water will be greater than T_l. It will be apparent that, since ions are differentially involved in exchange reactions, the ionic composition of drainage water may vary during the leaching process.

PECK and HURLE (1973) used Eq. (7) to estimate the characteristic time for equilibration of input and loss of Cl from previously forested catchments, which have been disturbed by development of farms. For a high rainfall area, they estimated T_l as about 30 years, but in a drier region where secondary salinity is extensive T_l ranged from 200 to 400 years. These times are in accord with empirical observations that in this region few areas of secondary salinity have decreased significantly. In fact, the overall area of affected soil is increasing, which is explained by the extension of areas of changed land use, and the lag between increased recharge and development of a water table close to the soil surface.

When changed land use significantly increases the rate of groundwater recharge, two values of T_l can be computed: that with the smaller initial recharge R_0, and that with the larger recharge R_1. It is appropriate to use values of R_1 when estimating the characteristic time for re-equilibration of the salt balance following the change, but the pristine value of T_l, calculated using R_0, is of interest too.

If the pristine value of T_l is small, say 100 years or less, then it is reasonable to assume that a balance existed between rates of input and loss of a mobile ion such as Cl. However, large values of T_l, perhaps greater than 1000 years, raise the possibility that even in the pristine state there was disequilibrium due to past climatic changes. From their estimates of the pristine rates of groundwater recharge, PECK and HURLE (1973) calculated values of T_l ranging from 30 to 7000 years; for a catchment with average rainfall 1100 mm/yr, T_l was 200 years. Thus the possibility of disequilibrium of the salt balance should not be overlooked, even in relatively high rainfall areas where man's disturbance of the environment has been minimal.

G. Summary

Secondary salinity affects soils both in irrigation and dryland farming areas. An associated, and in some cases more significant problem is increased salinity of surface water supplies from salt-affected watersheds.

The fundamental problem is disturbance of the local water balance in such a way that groundwater recharge is increased. Common forms of disturbance are irrigation and replacement of forest vegetation by crop or pasture species. Increased recharge disturbs the distribution of solutes within the soil-water system.

Characteristics of water movement in soils are such that there can be a long time lag between increased recharge and concentration of solutes at the soil surface, or their direct discharge into a stream. The time scale for approach to new water and salt balances is long too. It has been estimated to range up to 400 years in some farmed areas of south-western Australia.

References

BALDWIN, A. D.: Contribution of atmospheric chloride in water from selected coastal streams in central California. Water Res. Res. **7**, 1007–1012 (1971).

BETTENAY, E.: The salt lake systems and their associated aeolian features in the semi-arid regions of Western Australia. J. Soil Sci. **13**, 10–17 (1962).

BIGGAR, J. W., NIELSEN, D. R.: Miscible displacement and leaching phenomena. In: Irrigation of agricultural lands, Agronomy No. 11, pp. 254–274. Madison, Wisconsin: Am. Society of Agronomy 1967.

BITTINGER, M. W., TRELEASE, F. J.: Development and dissipation of a ground-water mound. Trans. Am. Soc. Agr. Eng. **8**, 103–106 (1965).

BONYTHON, C. W.: The salt of Lake Eyre—Its occurrence in Madigan Gulf and its possible origin. Trans. Roy. Soc. S. Aust. **79**, 66–92 (1956).

BOUGHTON, W. C.: Effects of land management on quantity and quality of available water: a review. Report No. 120, Water Research Laboratory, University of New South Wales, Australia (1970).

BROWN, P.: Water use and soil water depletion by dryland winter wheat as affected by nitrogen fertilization. Agron. J. **63**, 43–46 (1971).

CLARK, C.: Saline-seep development on non-irrigated cropland. In: Proceedings saline seep-fallow workshop. Bozeman, Montana: Cooperative Extension Service 1971.

CLEAVES, E. T., GODFREY, A. E., BRICKER, O. P.: Geochemical balance of a small watershed and its geomorphic implications. Geol. Soc. Am. Bull. **81**, 3015–3032 (1970).

ERIKSSON, E.: The yearly circulation of chloride and sulphur in nature: Meteorological, geochemical and pedological implications. Part II. Tellus **12**, 63–109 (1960).

ERIKSSON, E., KHUNAKASEM, V.: Chloride concentration in groundwater, recharge rate and deposition of chloride in the Israel Coastal Plain. J. Hydrol. **7**, 178–197 (1969).

FREEZE, R. A., WITHERSPOON, P. A.: Theoretical analysis of regional groundwater flow: 2. Effect of water table configuration and subsurface permeability variation. Water Res. Res. **3**, 623–634 (1967).

FREEZE, R.A., WITHERSPOON, P.A.: Theoretical analysis of regional groundwater flow: 3. Quantitative interpretations. Water Res. Res. **4**, 581–590 (1968).
FRIED, J.J., COMBARNOUS, M.A.: Dispersion in porous media. Advanc. Hydrosci. **7**, 169–282 (1971).
GARDNER, W.R.: Dynamic aspects of water availability to plants. Soil Sci. **89**, 63–73 (1960).
GARDNER, W.R.: Movement of nitrogen in soil. In: Soil nitrogen, Agronomy No. 10, pp. 550–572. Madison, Wisconsin: Am. Soc. Agronomy 1965.
GARDNER, W.R.: Water uptake and salt distribution patterns in saline soils. In: Isotope and radiation techniques in soil physics and irrigation studies, pp. 335–340. Vienna: International Atomic Energy Agency 1967.
GREENLEE, G.M., PAWLUK, S., BOWSER, W.E.: Occurrence of soil salinity in the drylands of southwestern Alberta. Can. J. Soil Sci. **48**, 65–75 (1968).
HADAS, A., HILLEL, D.: Steady state evaporation through non-homogeneous soils from a shallow water table. Soil Sci. **113**, 65–73 (1972).
HILLEL, D.: Soil and water: physical principles and processes. New York: Academic Press 1971.
HOLMES, J.W.: Salinity and the hydrologic cycle. In: Salinity and water use, pp. 25–40. London: Macmillan 1971.
JACOBSEN, T., ADAMS, R.M.: Salt and silt in ancient Mesopotamian agriculture. Science **128**, 1251–1258 (1958).
KUNZE, R.J., UEHARA, G., GRAHAM, K.: Factors important in the calculation of hydraulic conductivity. Soil Sci. Soc. Am. Proc. **32**, 760–765 (1968).
LIGHTFOOT, L.C., SMITH, S.T., MALCOLM, C.V.: Salt land survey, 1962: Report of a survey of soil salinity. J. Agric. West. Aust. **5**, 396–410 (1964).
MULQUEEN, J., KIRKHAM, D.: Leaching of a surface layer of sodium chloride into tile drains in a sand-tank model. Soil Sci. Soc. Am. Proc. **36**, 3–9 (1972).
NORTHCOTE, K.H., SKENE, J.K.M.: Australian soils with saline and sodic properties. Soil publication No. 27. CSIRO, Australia (1972).
PASSIOURA, J.B., ROSE, D.A.: Hydrodynamic dispersion in aggregated media, 2. Soil Sci. **111**, 345–351 (1971).
PECK, A.J.: Analysis of multidimensional leaching. Soil Sci. Soc. Am. Proc. **37**, 320 (1973).
PECK, A.J., HURLE, D.H.: Chloride balance of some farmed and forested catchments in southwestern Australia. Water Res. Res. **9**, 648–657 (1973).
PHILIP, J.R.: Theory of infiltration. Advanc. Hydrosci. **5**, 215–295 (1969).
PHILIP, J.R.: Flow in porous media. Ann. Rev. Fluid Mech. **2**, 177–204 (1970).
REEVE, R.C., FIREMAN, M.: Salt problems in relation to irrigation. In: Irrigation of agricultural lands, Agronomy No. 11, pp. 988–1008. Madison, Wisconsin: Am. Soc. Agronomy 1967.
SCHWARTZ, F.W., DOMENICO, P.A.: Simulation of hydrochemical patterns in regional groundwater flow. Water Res. Res. **9**, 707–720 (1973).
SCOFIELD, C.S.: Salt balance in irrigated areas. J. Agr. Res. (Washington) **61**, 17–39 (1940).
TALSMA, T.: The control of saline groundwater. Mededelingen van de Landbouwhogeschool, Wageningen, Netherlands **63**, 1–68 (1963).
TOTH, J.: A theoretical analysis of groundwater flow in small drainage basins. J. Geophys. Res. **68**, 4796–4812 (1963).
TSENG, M.T., RAGAN, R.M.: Behavior of groundwater flow subject to time-varying recharge. Water Res. Res. **9**, 734–742 (1973).
URIE, D.H.: Estimated groundwater yield following strip cutting in pine plantations. Water Res. Res. **7**, 1497–1510 (1971).
WOOD, W.E.: Increase of salt in soil and streams following the destruction of the native vegetation. J. Roy. Soc. West. Aust. **10**, 35–47 (1924).

Intermediary Remarks

A. Poljakoff-Mayber and J. Gale

The worldwide occurrence of salinity has been described in previous chapters. On every continent, and in practically every country, very extensive areas of soil are affected by primary, natural, or by secondary, often man-made, salinity. The excessive amounts of salts present in saline soils do not usually contain equal quantities of all the ions present. The dominant salts are usually NaCl, Na_2SO_4 or a mixture of the two. Sometimes large quantities of carbonates are present. The excess salts, especially in alkali soils, cause physical and chemical changes in soil structure, such as swelling of soil particles, which interfere with drainage and aeration (Chapter 3). The quality of water used for irrigation, the nature of the ground waters and the amount of annual precipitation are all factors which contribute to the properties of the substrate from which the plant draws its mineral nutrients (Chapters 2–4).

The natural environment, undisturbed by man, has through the millennia since the formation of soil, attained equilibrium. Maintenance of such an equilibrium depends to a very large extent on whether or not a favorable, hydrological balance can be maintained (Chapter 5). Irrigation and fertilization practices usually interfere with such a balance and eventually affect, not only the soil in question, but also the quality of the ground water, both in the irrigated area itself and also in more distant places. Not only mismanagement of irrigation, but also man-produced changes in natural vegetation, such as substitution of annual crops or pastures for forest areas, may affect the water balance and induce salinization (Chapter 5). Amelioration of the situation requires special agrotechnical practices such as drainage, leaching, controlled irrigation, special planting techniques and the choice of salt tolerant crops (Chapters 2 and 3).

Saline conditions drastically change the environment of the root—aeration, osmotic potential of the soil solution and the normal equilibrium of the dissolved ions. Mineral nutrition of plants in general and under saline conditions in particular has recently been reviewed by Epstein (1972) and by Rains (1972). Waisel (1962, 1972) also deals with this problem with reference to halophytes.

To some extent plants can adapt to the low osmotic potentials prevailing in saline soils by absorbing ions. As NaCl is the most common salt present in saline soils, most research on salinity has been done using NaCl. Both sodium and chloride are readily taken up by the root system of both halophytes and glycophytes. Most halophytes absorb sodium, transport it to the leaves and either excrete it through salt glands or tolerate it until the concentration is too high and the cells die. Plants can rid themselves of excess salt by shedding, yearly, the old, salt rich leaves.

In many glycophytes only very small amounts of sodium reach the leaves, most of it being retained in the roots or in the stems (Jacoby, 1964, 1965; LaHaye and Epstein, 1969, 1970). However, in other plant species, cations are absorbed in excess of anions and electrical balance is achieved by accumulation of organic acids (De Wit et al., 1963).

Calcium has long been known to have an ameliorating effect on the growth of plants under saline conditions (HYDER and GREENWAY, 1965; DEO and KANWAR, 1969, and see also BERNSTEIN, 1970; EPSTEIN, 1972, for older references). The effect has often been ascribed to Ca^{++} preventing the uptake of the toxic Na^+ ion, while allowing continued uptake of K^+ (e.g. WAISEL, 1962). Calcium ion also has a highly desirable flocculating effect on soil, whose aggregate structure has been dispersed by monovalent sodium. In this respect it should be remembered that much of the deleterious action of salinity on plants can be traced to its effects on soil. These may be chemical and physical (Chapter 3) or microbiological (e.g. SUBBA-RAO et al., 1972).

Recently LAHAYE and EPSTEIN (1969, 1970) demonstrated again the large effect that Ca^{++} ion may have on plants growing in salinized culture solutions. Working with beans *(Phaseolus vulgaris)* they found that 50 mM NaCl severely repressed growth, but growth was restored by the addition of 1 mM or more $CaSO_4$. At <0.1 mM $CaSO_4$ the sodium content of the leaves was 3.2 mg/100 mg dry weight, but dropped to 0.2 mg/100 mg when 3.0 mM $CaSO_4$ was added to the nutrient solution. They believe that the site of the Ca induced exclusion of Na^+ is the plasmalemma of the root cells. Results from one of their experiments showing a very dramatic effect of Ca^{++} can be seen in Fig. 1.

CHIMIKLIS and KARLANDER (1973) describe an interesting effect of Ca^{++} on salinized chlorella. Generally, salt was found to increase O_2 evolution and decrease CO_2 fixation in illuminated chlorella. They concluded that due to salinity more photosynthetic energy was expended on the exclusion of Na^+ and accumu-

Fig. 1. Appearance of bean plants after a 7-day period of growth in aerated nutrient solutions in the greenhouse. The concentration of NaCl was 50 mM, and that of $CaSO_4$ was (from left to right) zero, 0.1, 0.3, 1.0, 3.0, and 10 mM. From LA HAYE and EPSTEIN (1969) by kind permission of the authors and the AAAS

lation of K^+ and Cl^-, leaving less energy available for reduction of CO_2. Addition of Ca^{++} reversed this trend and increased the tolerance level to NaCl.

There is therefore no doubt that Ca^{++} ion has an ameliorating effect on plant growth under conditions of salinity. The mechanism of its action is quite clear in the soil, but, as demonstrated in Fig. 1, it also increases plant resistance to salinity under soilless conditions, where the mechanism for its action is far less certain.

Potassium has long been recognised as one of the essential macronutrients of plants. However sodium was not, until recently, considered to be an essential element. Halophytes tolerate very high concentrations of sodium, and it was shown that *Atriplex vesicaria* and several other species of *Atriplex* do require sodium, but only as a micronutrient (BROWNELL, 1968).

In saline soils the Na/K ratio is very high. It seems therefore logical to expect that halophytes and salt tolerant plants would have developed a mechanism for the preferential uptake of potassium from mixtures rich in sodium—i.e.—they should have a very developed "absorption System I" (EPSTEIN, 1972). RAINS and EPSTEIN (1967) have indeed shown that in *Avicennia marina* even absorption System II (in which excess sodium usually inhibits potassium uptake) absorbed potassium preferentially. Moreover, the presence of NaCl even caused some increase in potassium uptake.

Ion uptake by plant tissue is accompanied by increased respiration ("salt respiration"). RAINS (1972) sees no plausible explanation for this increase of respiration at salt concentrations at which absorption System II is activated. He discusses the alternative theories of energy supply for ion absorption by either the chemi-osmotic proton pump mechanism, in which anions move against an electrochemical potential gradient (e.g. PALLAGHY, 1969), or by the direct utilization of energy derived from the ATP produced in photosynthesis and respiration (see Chapter 8). As discussed in Chapter 9, the necessity to expend energy on moving salts against gradients is considered to be one of the main factors causing reduction of growth under saline conditions.

Salinity has also been implicated in the induction of micronutrient deficiencies. However, recent work of MAAS et al. (1972) produced no evidence of salt induced deficiencies of Fe, Zn or Mn in tomato, soya, beans or squash.

References

BERNSTEIN, L.: Calcium and salt tolerance of plants. Science **167**, 1387 (1970).

BROWNELL, P. F.: Sodium as an essential micronutrient element for some higher plants. Plant and Soil **28**, 161–164 (1968).

CHIMIKLIS, P. E., KARLANDER, E. P.: Light and calcium interactions in chlorella inhibited by sodium chloride. Plant Physiol. **51**, 48–56 (1973).

DEO, R., KANWAR, J. S.: Effect of saline irrigation waters on the growth and chemical composition of wheat. J. Indian. Soc. Soil Sci. **16**, 365–370 (1969).

DE WIT, C. T., DIJKSHOORN, W., NOGGLE, J. C.: Ionic balance and growth of plants. Versl. Landbouwk. Onderz. **69**, 15 (1963).

EPSTEIN, E.: Mineral nutrition of Plants: Principles and perspectives. New York: John Wiley and Sons 1972.

HYDER, S. Z., GREENWAY, H.: Effect of Ca^{++} on plant sensitivity to high NaCl concentration. Plant and Soil **23**, 258–260 (1965).

JACOBY, B.: Function of bean roots and stems in sodium retention. Plant Physiol. **39**, 445–449 (1964).
JACOBY, B.: Sodium retention in exeised bean stem. Physiol. Plantarum **18**, 730–739 (1965).
LA HAYE, P. A., EPSTEIN, E.: Salt toleration by plants. Enhancement with calcium. Science **166**, 395–396 (1969); **167**, 1388 (1970).
MAAS, E. V., OGATA, G., GARBER, M. J.: Influence of salinity on Fe, Mn, and Zn uptake by plants. Agron. J. **64**, 793–795 (1972).
PALLAGHY, C. K.: Salt relations of *Atriplex* leaves. In: R. JONES (Ed.): The biology of *Atriplex*, pp. 57–62. CSIRO Australia (1969).
RAINS, D. W.: Salt transport by plants in relation to salinity. Ann. Rev. Plant. Physiol. **23**, 367–388 (1972).
RAINS, D. W., EPSTEIN, E.: Preferential absorption of potassium by leaf tissue of the mangrove *Avicennia marina:* An aspect of halophytic competence in coping with salt. Australian J. Biol. Sci. **20**, 847–857 (1967).
SUDBA-RAO, N. S., LAKSHMI-KUMARI, M., SINGH, S., MAGU, S. P.: Nodulation of lucerne (*Medicago sativa* L.) under the influence of sodium chloride. Indian J. Agric. Sc. **42**, 384–386 (1972).
WAISEL, Y.: The effect of calcium on the uptake of monovalent ions by excised barley roots. Physiol. Plant. **15**, 709–724 (1962).
WAISEL, Y.: Biology of Halophytes. New York: Academic Press 1972.

Response of Plants to Salinity

Chapter VI

Morphological and Anatomical Changes in Plants as a Response to Salinity Stress

A. Poljakoff-Mayber

A. Introduction

Salinity is known to affect many aspects of the metabolism of plants and to induce changes in their anatomy and morphology. These changes are often considered to be adaptations which increase the chances of the plant to endure the stress imposed by salinity; alternatively, they may be considered to be signs of damage and disruption of the normal equilibrium of life processes.

The anatomical and morphological features typical of halophytes are usually considered to be adaptations to salinity. There is very little experimental evidence to show whether the same features occur when the halophytes are not exposed to the effects of salinity. Some investigations in this direction were carried out with ecotypes of the same species by Waisel (1972) and by Goodman (1973).

More evidence is available from glycophytes and especially cultivated crops, enabling comparison of the same features under saline and non-saline conditions. However, whether the changes occurring due to salinity in these plants should be considered as adaptations or as signs of damage, is open to discussion.

Salinity has been shown to affect the time and rate of germination, the size of plants, branching and leaf size, and overall plant anatomy. Succulence is one of the most common features of halophytes and it also occurs in many glycophytes after exposure to salinity for appreciable periods of time. It is often considered to be an adaptation which reduces the internal salt concentration. However, many halophytes apparently reduce their internal salt content with the aid of salt glands (a feature discussed in detail in the following chapter) or by salt exclusion by roots, as in some mangroves.

Strogonov (1962) has reviewed the research carried out on the effect of salinity, mainly in the USSR, for the period 1920–1960. He emphasizes the importance of combined anatomical-physiological studies for a better understanding of the affects of salinity on plants.

More recently, Waisel (1972) reviewed the various effects of salinity on plants, including the morphological and anatomical changes occurring in response to salinity, or typical for halophytes. Both authors (Strogonov, 1962 and Waisel, 1972) ascribe to salinity the induction of numerous structural changes:

increase of succulence;
changes in number and size of stomata;
thickening of the cuticle;
extensive development of tyloses;
earlier occurrence of lignification;
inhibition of differentiation;
changes in diameter and number of xylem vessels.

The information available is, however, very fragmentary and scattered over a wide variety of plants. It is practically impossible to draw a complete picture for any one of the species investigated.

B. Effect of Salinity on Growth

Salinity causes stunting of glycophytes. In contrast halophytes, although able to grow in a non-saline substrate, will usually grow better in presence of some salt. However, in experiments with the halophyte *Atriplex halimus* L. when grown in culture media salinized with NaCl, to different levels, under two different air humidities (all other climatic conditions e.g. temperature and light regime being equal) a different picture emerged (GALE et al., 1970 and unpublished results): When the humidity of the air was high, addition of NaCl to the growth medium depressed the growth of the plants (as expressed by the relative growth rate). When the humidity of the air was low, an optimal growth curve was obtained; the optimal NaCl concentration being 120 mM in addition to the nutrient medium (to give a total osmotic potential of -5.25 atm). At higher and lower NaCl concentrations growth rate was reduced (Fig. 1). Plant height responded in a similar manner; plants grown in the humid chamber with no NaCl in the growth medium were the tallest, ~60 cm, and their height was markedly reduced by addition of NaCl. If however, the plants were grown in the dry chamber, the height of the plants grown in absence of NaCl was only 25 cm, while in the optimal NaCl concentration (120 mM) it was about 47 cm, equal to that of the plants grown in the humid chamber at the same concentration of NaCl. Higher NaCl concentrations affected the plants in both chambers similarly. It seems therefore that salinity induces some stunting in halophytes, at least in *A. halimus*.

Although the plants grown in 480 mM NaCl, in both chambers were stunted—only 25 cm high—the plants in the dry chamber were much more leafy than those in the humid chamber. The dry weight ratio of leaves to stem in the dry chamber was 4, while in the humid chamber this ratio was only 3. The effect of

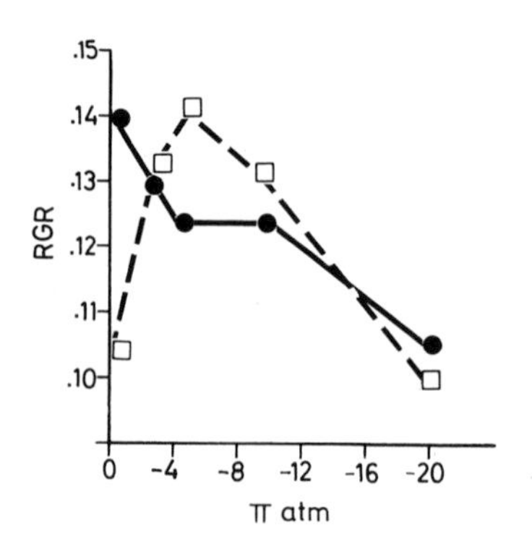

Fig. 1. Relative growth rate of *Atriplex halimus* plants grown at various salinity levels and under either dry (27% RH) or humid (63% RH) atmospheric conditions. ●—● Humid conditions. □—□ Dry conditions. Compiled from results of GALE et al. (1968)

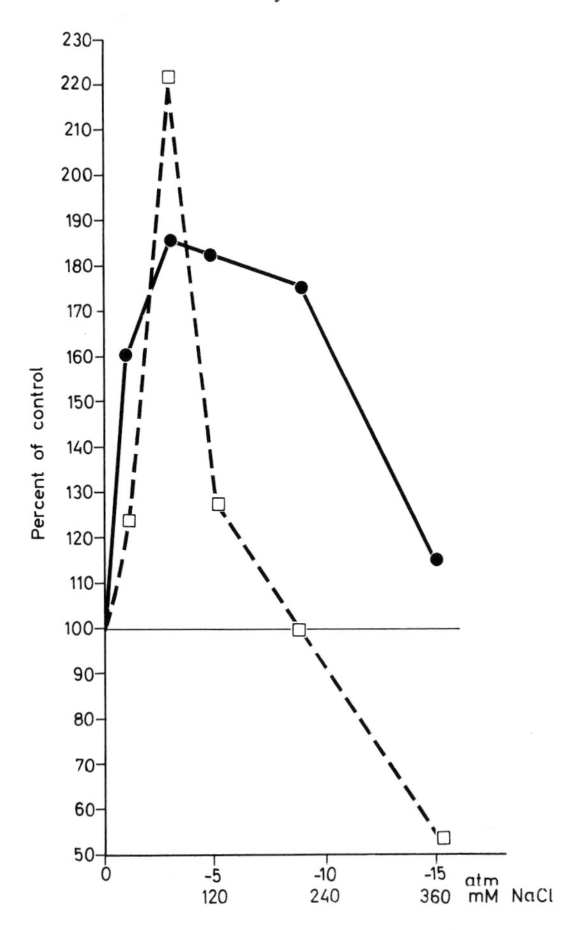

Fig. 2. Effect of salinity on leaf area of *Atriplex halimus*—Leaf area per plant. Plants grown in media salinized to different levels with either NaCl or Na_2SO_4. Results given as percent of control—plants grown in non-salinized medium. NaCl ●—●, Na_2SO_4 □—□. Results taken from GALE and POLJAKOFF-MAYBER (1970)

salinity on leafiness is shown also in Fig. 2, where the leaf area per plant is shown as related to various levels of either NaCl or Na_2SO_4 salinity.

These data indicate that halophytes grown in the absence of salt develop an "etiolated" appearance (Fig. 3): Little branching, small leaves and small total leaf area per plant. NaCl salinity induces branching and a large leaf area—which is also a large photosynthetic area. This effect is apparent at salinity levels at which the maximal RGR is also achieved.

From the results in Fig. 2 it also follows that Na_2SO_4 is much more damaging than NaCl for the growth of *A. halimus* and only a very narrow range of concentrations is favorable for development of an appreciable leaf area per plant.

Leaf area in bean plants decreased by approximately 20–40%, when grown in a substrate salinized to -3 atm NaCl (MEIRI and POLJAKOFF-MAYBER, 1970). The younger the leaf the greater was the effect of salinity. The surface of juvenile leaves

Fig. 3 A

Fig. 3 B

Fig. 3A and B. Plants of *Atriplex halimus* grown in absence (A) and in presence (B) of 300 mM NaCl. Taken from M. Sc. Thesis of SHAHAM-GOLD, 1963

was 10% smaller, the first trifoliate 30% and the 4th trifoliate 90% smaller, than the corresponding controls.

STROGONOV (1962) reports 50% inhibition of growth in tomatoes grown in soil containing 0.1% (of dry weight) chloride. The weight of fruit per plant was reduced by 90%. BLUMENTHAL-GOLDSCHMIDT and POLJAKOFF-MAYBER (unpublished results) noticed that at concentrations as low as -1 atm. NaCl the inter-vein tissue of tomato leaves grew more than the vein system and as a result, under saline conditions the leaves were very wavy and rippled.

C. Effect of Salinity on Succulence and on Leaf Anatomy

STROGONOV cited BATALIN (1875) as the first investigator to note the induction of succulence in *Salicornia herbacea* grown in media salinized with NaCl. The phenomenon did not occur if the plants were grown in non-salinized substrate, or in substrate salinized with $MgSO_4$, and was specific for NaCl. The succulence was due to development of larger cells in the spongy mesophyll and the presence of a multilayer palisade tissue which was absent in leaves of plants grown in nonsaline substrate. BEIKENBACH (1932) also noted thickening of palisade layers in *Aster tripolium* but he also reported thickening of the cortex parenchyma. STROGONOV (1962) himself carried out his experiments using mixtures of NaCl-Na_2SO_4 in the media. He considers this to be more typical of natural conditions. He studied the

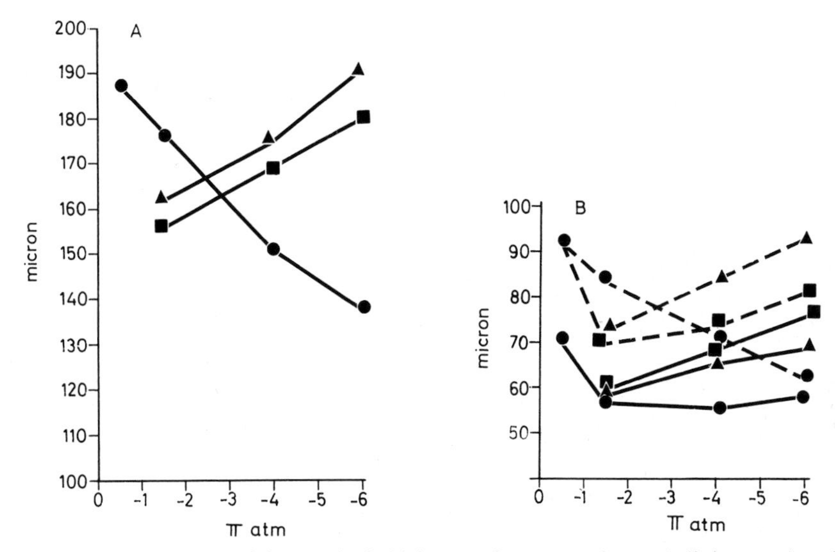

Fig. 4A and B. Effect of salinity on leaf thickness of tomato plants. Salinity produced by proportional increase of concentrations of all constituents of the basal medium (●), or by addition of NaCl (■) or Na_2SO_4 (▲) to the basal medium. (A) laminar thickness; (B) palisade and spongy layer thickness. Full line—palisade; Broken line—spongy layer. Compiled from data of HAYWARD and LONG, 1941

responses of cotton, tomatoes, sunflower, barley, beans and *Salicornia herbacea*. From his experiments he concluded that the presence of NaCl in the substrate caused succulence in tomato, cotton and *Salicornia;* while excess of Na_2SO_4 caused haloxeromorphism in cotton and only slight succulence in *Salicornia*. Marked xeric features appeared in *Salicornia* plants grown in non-saline substrate. He reports that Cl ions induced xeromorphism in barley.

HAYWARD and LONG (1941) also studied the effect of salinity on tomato plants. The normal leaf blade of tomato contains one layer of palisade cells and 4–5 rows of spongy mesophyll cells. The main vein projects from the lower surface of the leaf and its bilateral bundle is reinforced by abaxial and adaxial strands of collenchyma. HAYWARD and LONG subjected the plants to three types of salinity: proportional increase of all the constituents of the basal nutrient medium; salinization with NaCl in addition to the basal nutrient medium; and salinization with Na_2SO_4. With increasing equiosmotic concentrations in the base nutrient the thickness of the lamina decreased, while in NaCl and Na_2SO_4 the thickness of the lamina increased, more or less to the same extent (Fig. 4). If the stress was induced by the basic medium, the thickness of the spongy mesophyll decreased with increasing salinity, while thickness of the palisade layer decreased markedly with increase of salinity from -0.5 to -1.5 atm. and then remained constant with concentrations up to -6 atm. Under NaCl and Na_2SO_4 salinities, the thickness of both layers increased with increasing salinity, but Na_2SO_4 had more effect on the spongy mesophyll layer while NaCl had more effect on the palisade layer.

It is interesting to note that according to these data (Fig. 4) the increase in thickness caused by the high concentrations of NaCl or Na_2SO_4 has in no case exceeded the thickness of the control plants. Plants grown in -1.5 atm. salinity of all kinds had significantly thinner laminae than the control plants, and only at −6 atm. NaCl or Na_2SO_4 was the thickness of the control leaves achieved again. STROGONOV (1962) however, brings data showing doubling of the thickness of the leaf and palisade layer, due to chloride salinity. These data demonstrate the variability in the response of different plants to salinity, the complexity of the effect and its dependence on external conditions and on the nature of salinity.

Under natural conditions salinity is usually not at a constant level; it tends to increase between irrigations, or during periods without rain, and then suddenly decreases due to irrigation or rainfall. MEIRI and POLJAKOFF-MAYBER studied the effect of changing chloride salinity on growth of bean plants (MEIRI and POLJAKOFF-MAYBER, 1970; MEIRI et al., 1971). They found that the results for changing salinity regimes always fell between the results for the control plants and those exposed to continuous affect of salinity at its maximal level. It seems therefore that the total time of exposure to salinity at different levels is also important in assessing the effect of salinity on growth and morphology. The latter includes number of internodes, number of leaves, size of leaves, etc.

All the above data demonstrate the variability in the response of different plants to salinity, and it is therefore very difficult to define whether there are special responses typical to the halophytic or the glycophytic plants. POKROVSKAYA (1954 and 1957, as cited by STROGONOV, 1961) found that in halophytes cell division was inhibited by salinity, but cell extension was practically unaffected. This caused the development of typically large cells and leaf succulence. In glycophytes, exposed to the effect of salinity in the growth substrate, both cell division and cell extension were inhibited. She was unable, therefore, to offer an explanation for the development of succulence in these plants.

MEIRI and POLJAKOFF-MAYBER (1967) found that in leaves of bean plants exposed to salinity, both growth in area (which to a certain extent depends on cell division) and increase in thickness were affected. Growth in area of the juvenile leaves decreased more than 50% immediately following salinization of the medium, and continued to decrease with time. The daily increase in thickness stopped immediately on exposure to salinity, but was restored to its original rate after 24 hours and maintained a higher rate than the control leaves throughout the experiment (Fig. 5). The total increase in thickness was approximately 25%—two thirds of this was due to the increase in size of the cells of the spongy mesophyll layer and one third due to increase in the palisade layer. The thickness of the epidermal cells hardly changed. As the number of cells per unit area, of both the upper and lower epidermis, was larger in the leaves of plants exposed to salinity, it seems that the expansion of epidermal cells was more affected by salinity than was cell division.

UDOVENKO et al. (1970) did not find any increase of succulence, in response to salinity, in wheat leaves (using a relatively salt tolerant variety—Federation), thus confirming the observations reported by STROGONOV (1962) for barley. They grew the plants in NaCl, Na_2SO_4 or polyethyleneglycol-6000, in equiosmotic concentrations ($\pi = -7.1$ atmospheres) and there was no significant difference between the three treatments. Sometimes, even a decrease in leaf thickness occurred. UDOVENKO et al. (1970) suggest the possibility of different responses to salinity in mono- and dicotyledonous plants.

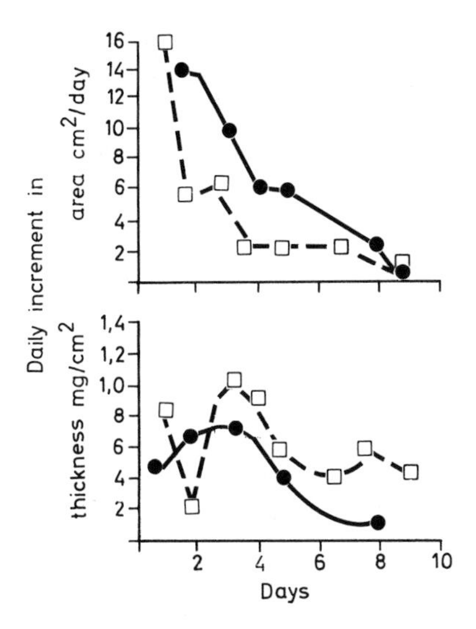

Fig. 5. Effect of salinity on growth in thickness and in area of primary leaves of bean plants. Graph compiled from data of MEIRI and POLJAKOFF-MAYBER (1967). Non saline control ●—●; grown in 72 mM ($\pi = -3$ atm) NaCl □—□

The changes in morphology described in the above experiments, although accompanied by thickening of the leaves are generally more typical of xeromorphism.

D. Effect of Salinity on Microscopic and Submicroscopic Structure of Leaf Cells

BLUMENTHAL-GOLDSCHMIDT and POLJAKOFF-MAYBER studied the submicroscopic changes in leaves of *Atriplex halimus* (1968) and a number of cultivated plants of differing salt tolerance (unpublished results). Studies on the leaf structure of some other *Atriplex* species were summarized by WEST (1970). Many species of the *Atriplex* genus are characterized by the so-called "Kranz" arrangement of the assimilatory tissue. This is also the case for *Atriplex halimus*.

According to FAHN (1974), *A. halimus* has an isolateral leaf structure. The epidermis is thin and covered by vesicular salt hairs. Underneath the epidermes of both sides of the leaf is a layer of water storing hypodermis cells. The mesophyll, or chlorenchyma, is made up of elongated "palisade" cells. The vascular bundles are surrounded by sheath cells which are extremely rich in chloroplasts. However, vacuoles are very scarce (BLUMENTHAL-GOLDSCHMIDT and POLJAKOFF-MAYBER, 1968). The sheath is open towards the abaxial side of the leaf. In leaves of plants grown in Knop's solution, the palisade cells contained long and thin chloroplasts, with comparatively few grana, many very long frets (intergranal lamellae) which were found to be continuations of the granal compartments. The chloroplasts in the bundle sheath cells were much shorter and thicker, the grana were much more abundant and had well defined margins. The frets were little developed. The bundle sheath cell chloroplasts contained numerous thick starch grains and many lipid droplets. Both the starch grains and the lipid droplets are much less common in the chloroplasts of palisade cells. Both types of chloroplasts have considerable amounts of stroma between the frets and large amounts of ribosomes. Both types of cells had large nuclei and numerous mitochondria. In the palisade cells the mitochondria were small, usually round, with numerous cristae. In the bundle sheath cells the mitochondria were even more abundant, comparatively large and often elongated. Endoplasmic reticulum was always present although not always very distinct. Plasmatic strands passing through the vacuoles were often found. Ribosomes and polysomes were abundant and were usually more distinct in the palisade than in the bundle sheath cells.

The changes induced by addition of NaCl to the growth medium became more distinct with increasing salinity and with prolongation of the period of exposure to salinity. The most obvious changes were observed in the chloroplasts. Swelling occurred at the granal and fret compartments. In chloroplasts of plants grown at -3 to -7 atm (72–168 mM NaCl) changes were hardly noticeable, but were very pronounced in plants grown at -11 to -19 atm. In plants grown at -19 to -23 atm a complete distortion of chloroplast structure occurred. Swelling of the compartments of the grana and of the frets has progressed to such an extent that the margins could not be distinguished and the chloroplasts seemed to be granaless.

The mitochondria in leaf cells of *Atriplex* plants grown under salinity (BLUMENTHAL-GOLDSCHMIDT and POLJAKOFF-MAYBER, 1968) appeared to be less electron dense and their cristae swollen. SHIMONY (1972) studied the mitochondria in the cells of the salt glands on leaves of *Avicennia marina* and *Tamarix aphylla*. She found that the mitochondria may appear either electron dense or transparent, depending on the secretory state of the gland. These changes are reversible. Transparent mitochondria apparently are active in cation absorption. She suggests that ion absorption may be selective and that the mitochondria retain ions whose loss would be against the overall strategy of the cell, but her evidence for this is inadequate.

The nuclei seemed to be unchanged except that the double membrane was swollen. The tonoplast showed many inward coils and folds and the plasmalemma appeared wavy and retracted from the cell wall. This feature of the plasmalemma became more pronounced with increasing salinity and was most evident after osmic acid vapor fixation (BLUMENTHAL-GOLDSCHMIDT and POLJAKOFF-MAYBER, 1968).

STROGONOV (1962) studied the mechanism of growth inhibition caused by salinity. He shows pictures of leaf cells from tomato, barley and cotton plants exposed to salinity. In all these cells the protoplasts retreated from the cell walls and the cells appeared to be plasmolyzed. He also reports that in these cells breakage of plasmodesmata occurred. In all these pictures the cells, from leaves exposed to NaCl salinity, look considerably bigger than the cells of the control leaves, except in barley, where they appear to be smaller. In chloride treated plants the intracellular spaces were considerably larger. Na_2SO_4 salinity caused much less damage, less plasmolysis and no severance of plasmodesmata. In *A. hal-*

imus (BLUMENTHAL-GOLDSCHMIDT and POLJAKOFF-MAYBER, 1968) Na_2SO_4 salinity caused changes similar to but less severe than those caused by NaCl.

It is interesting to note that no such submicroscopic changes were found in *Atriplex* leaves (less than one year old) taken from a plant in a natural saline near the Dead Sea. Only leaves from the previous year showed disruptions of the internal cellular structure.

Von WILLERT and KRAMER (1972) showed that NaCl treatment of *Mesembryanthemum* plants was accompanied by specific cytological changes which were coupled to a change from the Calvin pathway of CO_2 fixation to the crassulacean acid type metabolism. Extensive vacuolation appeared in the leaf cells and "vacuole-like" spaces occurred under the chloroplasts between the plasmalemma and the cell wall. In these spaces "Hecht-fibres" could be distinguished. They describe these fibres as single membranes with a structure of a unit membrane. They also report distortion of the inner structure of the chloroplasts showing rolled up lamelar system, although the granae were still distinguishable.

It seems therefore that the organelles most affected by salinity are the chloroplasts. BLUMENTHAL-GOLDSCHMIDT and POLJAKOFF-MAYBER (unpublished results) studied the effect of salinity on the chloroplast of tomatoes and spinach leaves. Both are considered to be relatively salt tolerant crops (RICHARDS, 1954). The chloroplasts of the tomato mesophyll cells contain one or more large starch grains, situated in the middle of the organelle. The starch grains are surrounded by a relatively thin membranal structure. The chloroplasts contained distinct grana with well defined margins and clear intergranal spaces with not too many frets. In chloroplast not containing starch grains the grana were very numerous. No lipid droplets could be observed and only little of the space was filled with matrix and ribosomes. After 10 days growth in a medium salinized to -1 atm NaCl, swelling could already be observed in the mesophyll chloroplasts, mainly in the granal compartments and especially at their edges, near the margins. After one week at -2 atm the distance between the frets increased markedly (Fig. 6A) but grana could still be distinguished. After 5 days at -4 atm NaCl the grana were hardly noticeable (Fig. 6B), the whole intralamellar system was swollen.

Spinach *(Spinacea oleracea)* is considered to be much more tolerant to salinity than tomatoes (RICHARDS, 1954). In agreement with this, BLUMENTHAL-GOLDSCHMIDT and POLJAKOFF-MAYBER found that the changes due to salinity occurred at higher NaCl concentrations and after longer exposures (π lower than -5 atm, exposure of 30 days and more). Basically however, the changes were the same: swelling withing the granal loculi and frets, accumulation of lipid droplets and eventually no grana could be identified. Spinach chloroplasts contain large amounts of matrix and numerous ribosomes. This was not affected by salinity. STROGONOV et al. (1970) described similar changes in chloroplasts in leaves of tomato and pea plants exposed to NaCl salinity. Salinization with Na_2SO_4, in tomatoes and spinach, caused essentially the same effects but at much higher concentrations and after much longer exposures. The final damage was also less severe.

In spite of the extensive ultrastructural changes in the chloroplasts of spinach, no clear cut data could be obtained for any disturbances in the light reactions of these chloroplasts (AMIR, 1968). However, the possibility cannot be excluded that

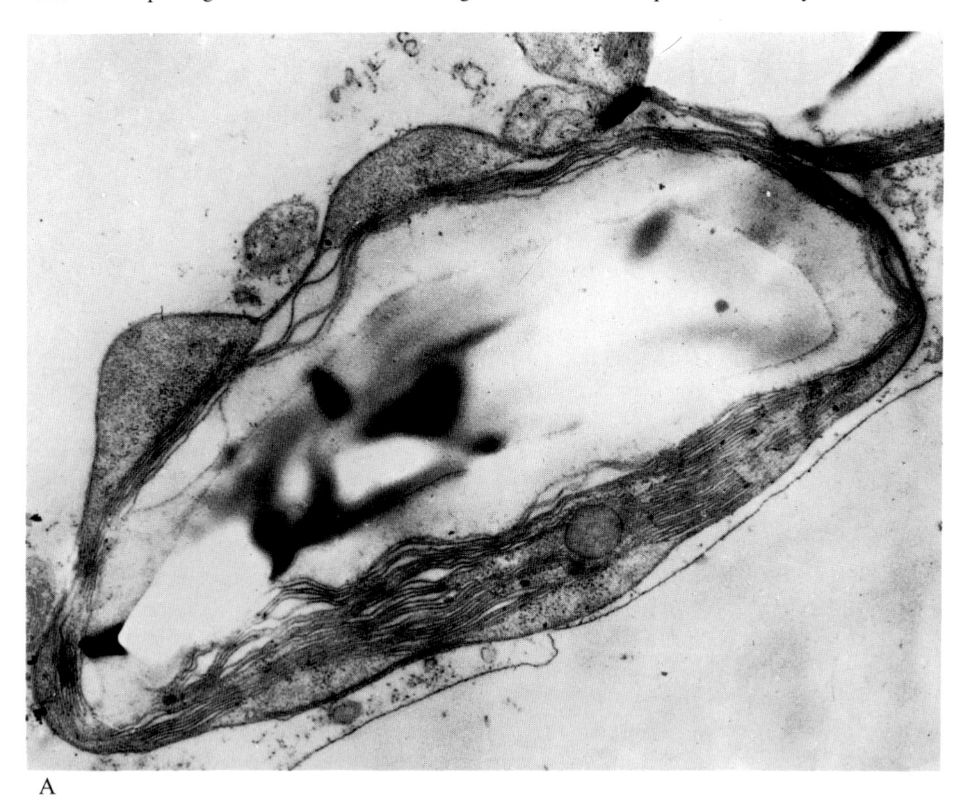

A

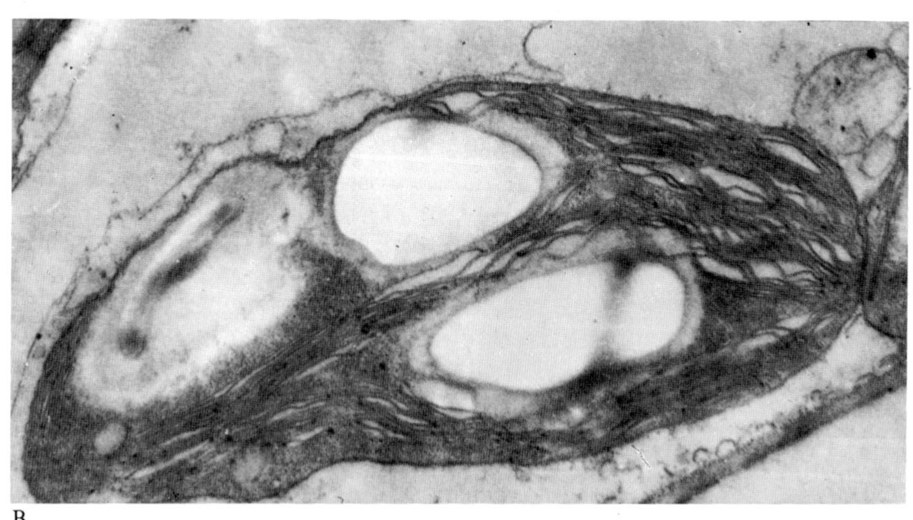

B

Fig. 6 A and B. Chloroplasts from mesophyll cells of tomato plants *(Lycopersicum esculentum var. Marmand)* grown in Knop's nutrient medium salinized with NaCl. (A) Plants were grown for 7 days in medium containing 24 meq/l NaCl (−1 atm). Fixation with 2% permanganate ×22600. (B) Plants were grown for 5 days at 96 meq/l NaCl (−4 atm). Fixation with 2% OsO_4 magnification ×17000. (BLUMENTHAL-GOLDSCHMIDT and POLJAKOFF-MAYBER, unpublished results)

the high variability in the experiments in which Hill reaction and photosynthetic phosphorylation were measured, masked the actual functional damage due to salinity.

Swelling of chloroplasts at low osmotic potentials was demonstrated also *in vitro*, with isolated chloroplasts (BLUMENTHAL-GOLDSCHMIDT and POLJAKOFF-MAYBER, 1966).

STROGONOV et al. (1972) found that NaCl induced shrinkage of the chloroplasts; the rate of photosynthetic phosphorylation increased and the rate of Hill reaction decreased with decreasing osmotic potential ("increased osmotic pressure"—in their words). Sodium sulphate, however, induced a decrease in photosynthetic phosphorylation. They believe that the effect is due to the anion and that the cation has practically no effect on these reactions.

E. Effect of Salinity on Stem Structure

HAYWARD and LONG (1941) studied the effect of salinity on the anatomical structure of tomato stem. A piece of stem was always taken from the median part of the basal internode immediately above the cotyledonary node.

The diameter of the stem, under saline conditions, was smaller than that of the control due mainly to reductions in the vascular tissue. The reduction in the parenchymatous tissue of the cortex and of pith was much smaller. The changes in size of the largest bundle in the transection (Fig. 7A) may serve as a criterion for

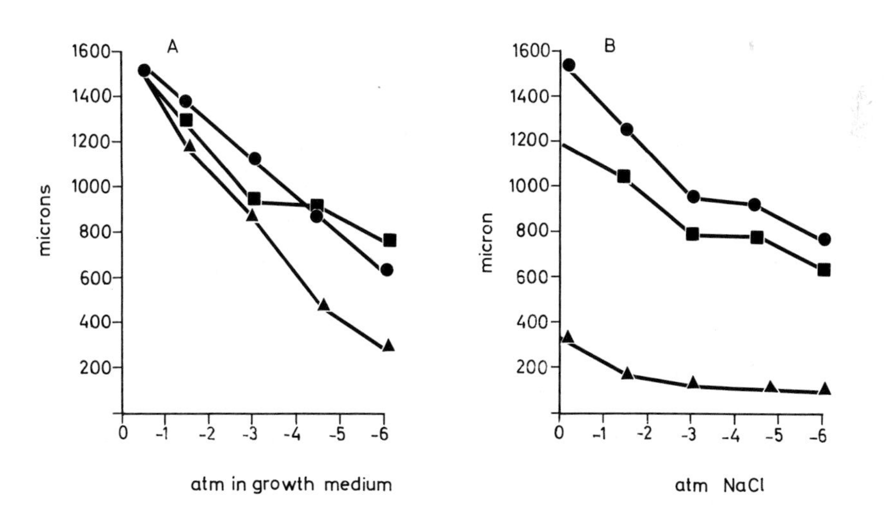

Fig. 7A and B. Changes in the amount of vascular tissue induced by salinity in tomato stem. (A) Effect of various types of salinity on the diameter of the largest bundle in the transverse section. ●—● Base nutrient medium in which concentration of all constituents was increased proportionally. ■—■ Osmotic potential achieved by addition of NaCl to the basic nutrient. ▲—▲ Osmotic potential achieved by addition of Na_2SO_4 to the basic nutrient. (B) Effect of NaCl on the diameter of the largest bundle (●), Xylem (■) and phloem (▲). Compiled from data of HAYWARD and LONG (1941)

the reduction in the amount of total vascular tissue induced by salinity. The diameter of the bundle decreased with increasing salinity: Sulphate salinity being the most damaging while the increased concentration of the basic medium was the least damaging. The reduction in the secondary xylem was parallel to the reduction in the diameter of the bundle (Fig. 7 B). Percentagewise the reduction in the secondary phloem was even larger. Cambial activity was reduced, in the presence of NaCl salinity, by 25% in media where $\pi = -4.5$, and by 34% where $\pi = -6$ atm. Na_2SO_4 salinity retarded cambial activity more than did NaCl (40% when $\pi = 4.5$ atm.). When $\pi = -6$ atm. no evidence was found of interfascicular activity. At this salt concentration cambial activity was 60% less than when $\pi = -1.5$ atm. Extensive accumulation of starch grains was observed in the endodermis and pith cells, and in the vascular paranchyma cells.

The above results of HAYWARD and LONG (1941) for tomato, are somewhat different from those of STROGONOV for cotton stems (1962). He worked with mixed chloride-sulphate types of salinity. When chloride was the dominant ion fewer tracheary elements developed and their diameter was bigger. When sulphate was the dominant ion the number of the elements in the stem was larger but they were of smaller diameter. On the whole, STROGONOV suggested that salinity inhibited differentiation in glycophytes and stimulated the increase in cortical parenchyma and pith tissue. STROGONOV also cited findings of GOREV (1954) showing excessive development of tyloses, due to salinity, in the xylem of grapevines and earlier lignification of branches as compared to vines growing in non-saline soil (PETROSYAN, 1960).

STROGONOV (1962) has also investigated the effect of salinity on the structure of stem tissue of *Salicornia herbacea*. The stem was much thinner, in the *absence* of NaCl, the cortical and pith tissues were less developed and water conducting tissue was very poorly developed. In presence of NaCl the development was more normal, all elements were more developed and the stem was thicker.

It seems that there is not enough descriptive evidence available to allow for generalization in describing the effect of salinity on the anatomical structure of the stem. The existing evidence is insufficient for differentiating between various responses of different plants to different types of salinity.

F. Effect of Salinity on Root Structure

GINZBURG (1964) made an intensive study of the interrelationships between root form, anatomy and the conditions of the habitat. He investigated numerous species collected in different habitats, from the Dead Sea salines and xeric desert regions to habitats having a relatively good water supply such as river beds. A survey of 30 different species showed that the compactness and degree of branching depend very much on the conditions of the "microhabitat" or the exact locality in the wider habitat. The roots of plants such as *Arthrocnemum glaucum*, *Suaeda monoica* and *Atriplex halimus* can reach to a depth of 4–5 m, the roots of plants such as *Prosopis farcta* and *Alhagi maurorum* can reach to a depth as great as 20 m.

Table 1. Comparison of some anatomical parameters of plants from different habitats. (Compiled from data of GINZBURG, 1964)

			Ratio between width of Casparian strips and radial wall	Number of cortex layers
A	Hydrohalophytes	*Suaeda monoica*	1.0	2.0
		Suaeda fruticosa	1.0	2.0
		Seidlitzia rosmarinus	1.0	3.8
		Statice pruinosa	1.0	2.0
		Nitraria retusa	1.0	2.5
		Arthrocnemum glaucum	1.0	2.9
B	Xerohalophytes	*Reaumuria palestina*	0.9	3.0
		Salsola baryosma	0.88	2.8
		Zygophylum dumosum	0.87	2.2
		Prosopis farcta	0.80	4.8
		Anabasis articulata	0.80	4.6
		Chenolea arabica	0.68	3.0
		Calligonum comosum	0.64	2.0
C	Sand dunes and river bed, Saharo-Sindian plants	*Atriplex halimus*	0.63	2.0
		Anabasis setipera	0.60	4.2
		Hamada salicornia	0.55	2.0
		Zilla spinosa	0.52	3.0
		Tamarix aphylla	0.52	8.6
		Retama roetam	0.50	4.8
		Artemisia herbaalba	0.50	4.6
D	Cultivated plants	*Phaseolus sp.*	0.33	14
		Linum	0.33	10
		Tomato	0.40	7
		Lettuce	0.27	6

Comparison of the anatomical features of species usually growing in arid or saline areas with those of glycophytes, suggested that the main differences may be found in the following three parameters:

1. Width of the Casparian strips on the radial walls of the endodermis.
2. Number of cell layers in the cortex.
3. The relative surface area of the tracheae in the transverse section of the root.

The ratio between the width of the Casparian strips to the width of the radial wall of the endodermis (Table 1) shows a decrease from 1 in the hydrohalophytes, to 0.9–0.8 in the xerohalophytes and to 0.6–0.5 in the dune and riverbed plants. In the hydrohalophytes (from flooded saline habitats) the whole of the radial wall was found to be covered with the Casparian strip. However, the number of cortical layers did not show any clear correlation to the conditions of the habitat.

The halophytes had a rather narrow cortex with only 2–3 layers of cells, while plants of the dunes and riverbeds had four or more layers. *Atriplex halimus*, although collected from a river bed (Table 1) showed a thin cortex like a real halophyte. *Prosopis farcta*, although collected in a saline environment, had a relatively wide cortex. An exceptionally wide cortex was found in *Tamarix aphyla*.

The ratio of tracheal area to the transverse section area of the root did not show any relationship to the conditions of the habitat.

The only parameter, therefore, that did show a significant correlation with salinity was the width of the Casparian strips. This is a fact which should be examined in greater detail in connection with the proposed pyhsiological role of the endodermis.

JONES and HODGKINSON (1970) studied the root distribution of two species of salt accumulating *Atriplex*—*A. nummularia* and *A. vesicaria*. Although both grow in very arid habitats, they show a low root/shoot ratio and not the high ratio so typical for xerophytes (OPPENHEIMER, 1960). The authors regard this feature as indicating that *Atriplex* originated in a saline habitat.

UDOVENKO et al. (1970a) report that NaCl, Na_2SO_4 and polyethyleneglycol 6000 at $\pi = -7.1$ atm, had practically identical effects on roots of wheat. All three induced thinner and less branched roots, decreased the width of the cortex and the diameter of the stele and brought about an increase in the thickness of the cell walls. The diameter of the tracheary elements was smaller than in the controls. In the two salt treatments, but not in carbowax, the passage cells in the endodermis disappeared.

UDOVENKO et al. (1970b) studied the effect of salinity (NaCl $\pi = -7.1$ atm) on the ultrastructure of the differentiation zone of roots of wheat (var. Federation and Diamant) and *Vicia faba*. After exposure of the tolerant wheat variety (Federation) to NaCl for 24 hrs, the mitochondria in the epidermis and pericycle cells increased in size from approx 0.5–0.7 μ to 0.6–0.9 μ. The density of the matrix decreased and the distance between the cristeae increased (i.e. the mitochondria had a swollen appearance), and the distance between the outer and inner membrane increased due to salinity. In the less tolerant variety the swelling was even more apparent, the cristeae appeared as inflated bubbles attached to the mitochondrial membrane. Similar observations were also made with bean roots.

After 5 days of exposure to salinity, the mitochondria of the tolerant wheat variety reverted to a normal appearance and after 10 days exposure the mitochondria in all three plants appeared normal—as if adaptation had occurred.

On short exposure to salinity the Golgi system also appeared swollen and small vacuoles containing dark material appeared near the edges of the cisternea, but again, after 10 days exposure to salinity everything reverted to normal. Similar swelling on short exposure and reversal to normal after long exposure to salinity, were observed in the endoplasmic reticulum and in the leucoplasts. No changes due to salinity were observed in the nuclei and there was no decrease in the number of polysomes.

STROGONOV et al. (1970) studied the changes due to salinity occurring in nuclei of cells of young leaves, tip and base of the stem and root tip of pea plants (variety "Inexhaustible"—as translated from Russian). They found small changes in volume of nuclei, but the ratio of the long to short axis of the elipsoid was not affected. However, they reported increase of the amount of DNA per cell (Table 2) and large variability in the ploidy. They assumed that it is the sodium accumulating in the cells that induced heteroploidy. The DNA content per cell was reported to vary between the equivalents of 2n and 16n. They cite BELL (1964) and claim that polyploidy is the response to "conditions unfavorable for normal life pro-

Table 2. Nuclear volume (μ^3) and relative content of DNA in cells of different organs of pea plants exposed to salinity. (Compiled from data of STROGONOV et al., 1970)

	Nuclear volume		DNA content	
	Control	Saline	Control	Saline
Leaf	19.7 ± 1.5	24.5 ± 3.6	2.03 ± 0.18	3.00 ± 0.15
Tip of stem	243.0 ± 34.0	194.0 ± 28.0	7.65 ± 1.13	10.50 ± 1.15
Base of stem	62.0 ± 7.2	60.3 ± 4.7	3.8 ± 0.48	6.03 ± 0.36
Tip of root	160.0 ± 11.0	180.0 ± 18.0	4.01 ± 0.23	6.45 ± 0.60

cesses". They claim that this polyploidy is reversible: it occurs during periods of stress (water stress for example) and reverts to normal with alleviation of stress. WULF (1937, as cited by STROGONOV et al., 1970) reports polymorphism in *Juncus buffonius* where, in the halophytic forms, the number of chromosomes reaches 120 while in the non-halophytic form it is only 20.

WITSCH and FLUEGEL (1951), as cited by STROGONOV et al., (1970) bring evidence from *Kalanchoe blassfeldiana*, which under short day conditions is succulent and has a 32n ploidy in the leaf cells. When this plant is transferred to long days the ploidy changes to 8n and the plant becomes non-succulent.

STROGONOV et al. (1970) consider the hetero-ploidy as an ontogenetic adaptation to salinity while polyploidy is a phylogenetic adaptation. However, at the same time they bring data showing that plants having 3n are more salt tolerant than those having 2n or 4n. They also show that adaptation to salinity is accompanied by increase in size and number of nucleoli, but that after prolonged exposure everything reverts to normal. The evidence for the data brought by STROGONOV et al. (1970) is not convincing, but it may be of interest to follow up some of their ideas and to test their validity.

NIR, HASSON-PORATH, and POLJAKOFF-MAYBER (unpublished results) studied the submicroscopic structure of root tips of peas and maize exposed to salinity for 10–15 days and of *Tamarix* after exposure for a month or more.

In pea root tips exposed to 120 mM ($\pi = -5$ atm) NaCl for 10–15 days, three things were very apparent as compared with the controls: the larger number of mitochondria per cell, the increase in the number of endoplasmic reticular elements and the swelling of the Golgi cisterneae. The increase in the number of mitochondria per cell was not established quantitatively, but there was a very clear impression of this from observations of very many electron microscopic fields taken from many root samples. This is a quite surprising finding, as the rate of respiration of pea root tips from plants grown in saline substrate was depressed.

In maize root tips, similar changes began to appear only after exposure to 240 mM NaCl. Lower salt concentrations, although depressing growth to some extent had no effect on submicroscopic structure. In maize exposed to 240 mM NaCl all the changes described for pea roots were observed and in addition a large number of lipid droplets accumulated in the cytoplasm. Similar accumulation of lipid droplets was described by NIR et al. (1970) in maize roots exposed to severe water stress. The droplets disappeared when roots were allowed to recover

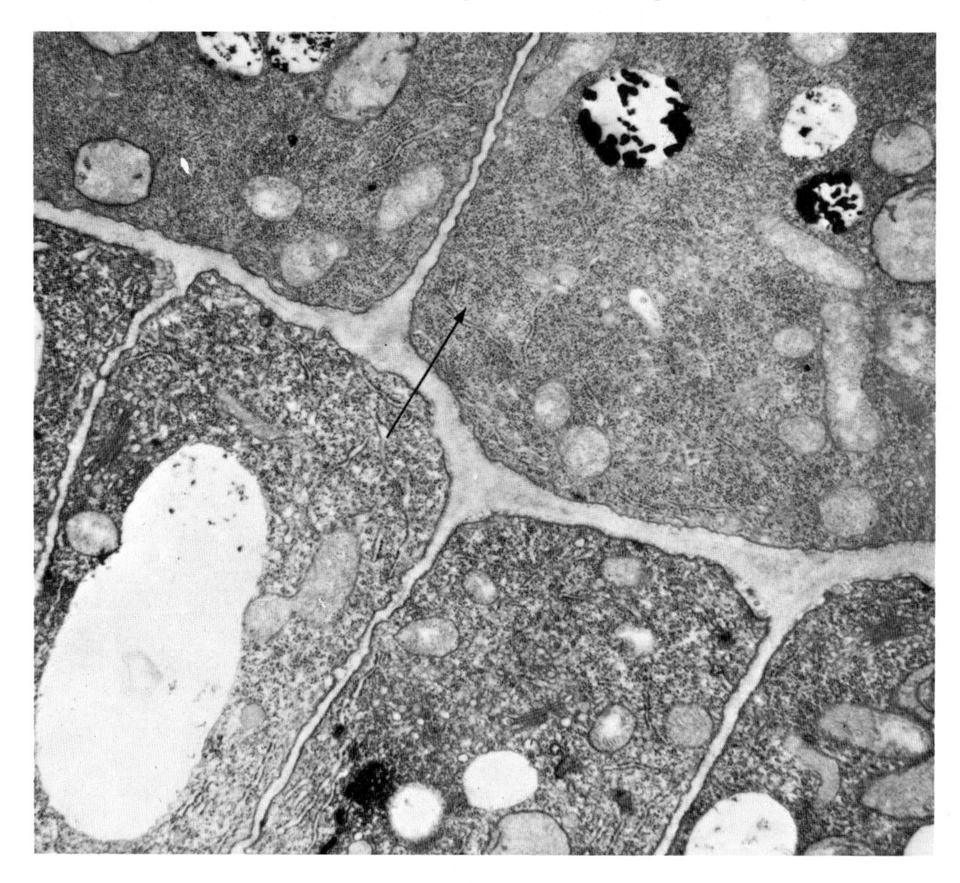

Fig. 8. Epidermal and subepidermal cells of *Tamarix* root tips from cuttings rooted in Hoagland's solution. Arrow shows towards the centre of the root. × 11250. (From NIR and POLJAKOFF-MAYBER, unpublished results)

from the water stress. It was not established whether a similar reversal occurs after returning the roots to non-saline media.

Tamarix roots have a distinct epidermal layer and several subepidermal layers of cells that appear to be different from those of the rest of the root. In these cells the cytoplasmic density seems to be very high and after staining with lead citrate they appear to be darker than the inner layers (Fig. 8). The reticular elements in these cells are very numerous and a little swollen, ribosomes are less numerous than in the inner cells. Very numerous Golgi elements are found in some of the cells. If roots are grown in media containing 288 mM NaCl the Golgi elements become hypertrophic and very vesicular (Fig. 9) and the mitochondrial matrix seems to be very light (see section on leaf mitochondria).

If cuttings rooted in 288 mM NaCl were promptly transferred to 480 mM NaCl, very drastic changes occurred in the epidermal and subepidermal cells. The cuttings could survive in such high salinities. Although growth practically

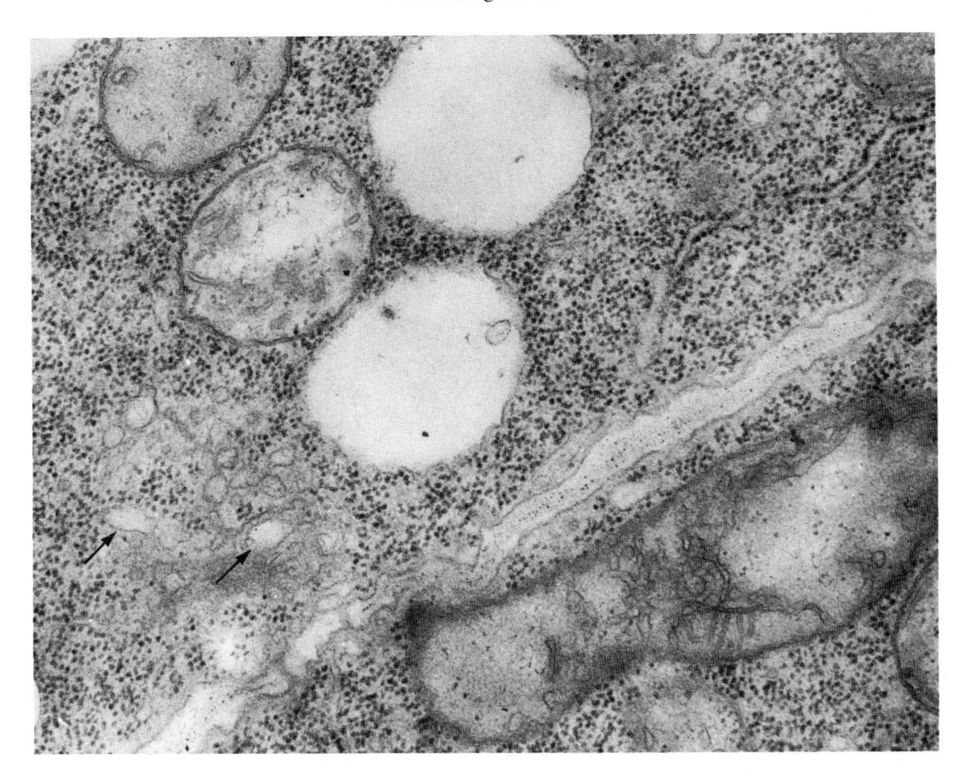

Fig. 9. A part of a subepidermal cell of *Tamarix* root grown at -12 atm NaCl. Arrows show typical strongly vesicular golgi bodies. $\times 27500$. (From NIR and POLJAKOFF-MAYBER, unpublished results)

stopped the roots did not look necrotic. However, the epidermal and subepidermal cells completely lost their internal structure and were filled with an electron dense precipitate. They were, apparently, dead (Fig. 10). Cytoplasm and organelles could be distinguished in the inner cells (Fig. 11), but a high degree of disorganization was observed and the nuclei were spread along the outermost wall of the cell. There is no functional explanation as yet, for these changes in *Tamarix* roots.

The most obvious changes induced by salinity in all the investigations described above are those taking place in the Golgi elements. This may have to do with the role of the root, as the ion absorbing organ, during the osmotic adaptation occurring under saline conditions.

G. Concluding Remarks

There is not yet enough evidence for interpreting the anatomical and submicroscopic changes induced by salinity, nor can the various findings be clearly correlated with distinct metabolic events. The observations summarized in this chapter clearly demonstrate that exposure to salinity during growth induces stunting of growth and structural changes at various levels of organization.

Fig. 10. Epidermal and subepidermal cells of *Tamarix* roots grown at −12 atm NaCl then transfered to −20 atm. × 3050. (From NIR and POLJAKOFF-MAYBER, unpublished results)

Stunting of growth is not necessarily always a disadvantage. Completion of the developmental cycle and production of a large number of seeds are the key processes for survival and continuation of the species. In salines of arid and semi arid regions, a large vegetative body with a large leaf area, although beneficial for photosynthesis, also exposes a large transpiring area. A small vegetative body and rapid maturation and flowering may be of great advantage. Low yields, (fresh and dry weights) although disadvantageous from an agricultural point of view, may ecologically-wise be a very useful strategy (GOODMAN, 1973).

The microscopic and submicroscopic changes, occurring in response to salinity, apparently vary in different plants, the division by no means falling between halophytes and glycophytes. Clearly, there is not yet enough evidence, on either anatomical or morphological changes induced by salinity, nor on changes in fine structure. The nature of these changes and their physiological and metabolic significance still await further investigations. This applies to both glycophytes and halophytes. It is as yet impossible to decide which of the changes induced by salinity are beneficial adaptations and which are signs of damage. Unfortunately we do not clearly understand which are the special properties that enable the plant to endure the hazards of a saline environment.

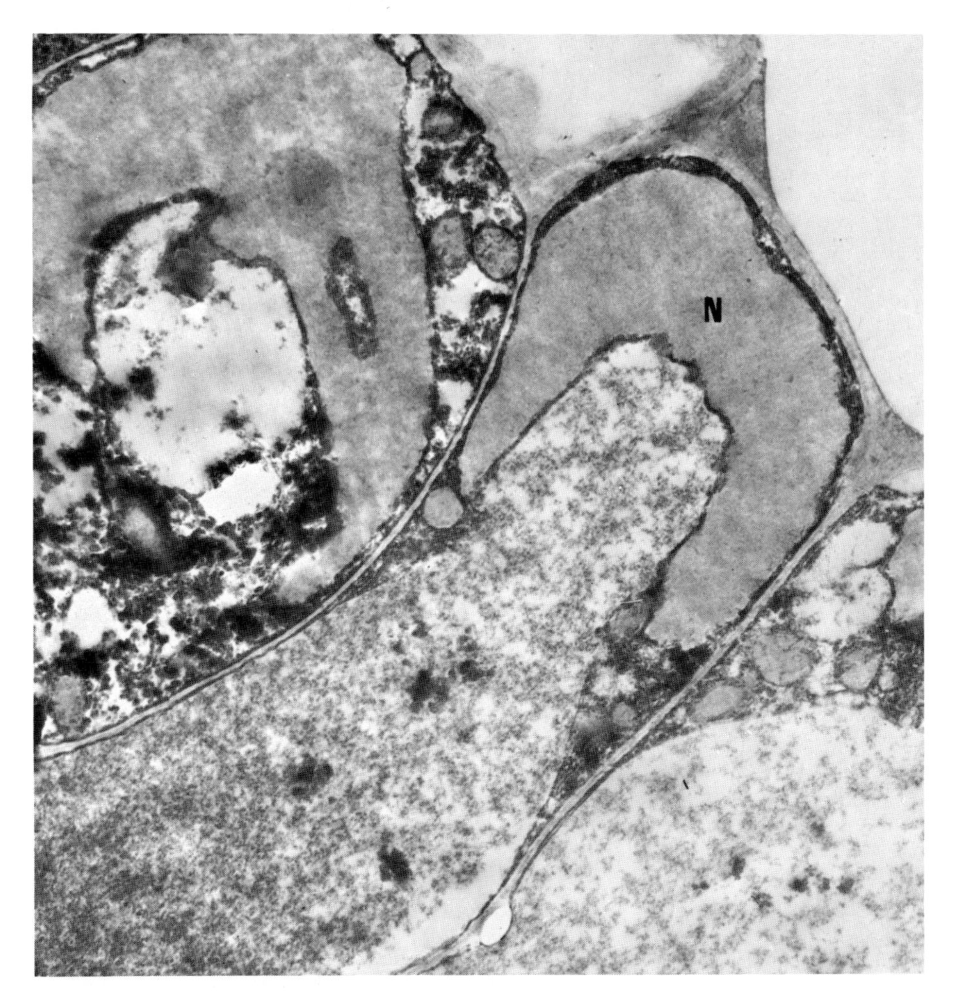

Fig. 11. An inner subepidermal layer from the root shown in Fig. 10. (*N*) Nucleus. ×12425. (From NIR and POLJAKOFF-MAYBER, unpublished results)

Acknowledgements

The unpublished information of BLUMENTHAL-GOLDSCHMIDT, NIR, HASSON-PORATH, and POLJAKOFF-MAYBER, cited in this chapter was obtained under a grant from the United States Department of Agriculture (A10-SWC-7; FG-Is-179) to POLJAKOFF-MAYBER and MEIRI.

References

AMIR, R.: Effect of salinity in the growth medium on activity of isolated chloroplasts. M. Sc. Thesis (in Hebrew), The Hebrew University of Jerusalem (1968).

BATALIN, A. F.: Isledovaniya nad vliyaniem khloristogo natriya na razvitie trekh solanchakovykh rastenii: *Spergularia media var. marganita*, *Salsola soda* i *Salsola mutica* (Effect of sodium chloride on the development of three solonchak plants). Trudy S. Pb. Obshehestvci estestvoisphytanii **6**, Protok. (1875).

BEIKENBACH, K.: Zur Anatomie und Physiologie einiger Strand- und Dünenpflanzen. Beiträge zum Halophytenproblem. Beitr. Biol. Pflanz. **19**, 113 (1932).

BELL, C. K.: Incidence of polyploidy correlated with ecological gradients. Evolution **18**, 510–511 (1964).

BLUMENTHAL-GOLDSCHMIDT, S., POLJAKOFF-MAYBER, A.: Comparison of the osmotic volume changes in chloroplasts isolated from different plants. Plant and Cell Physiol. **7**, 357–362 (1966).

BLUMENTHAL-GOLDSCHMIDT, S., POLJAKOFF-MAYBER, A.: Effect of salinity on growth and submicroscopic structure of leaf cells of *Atriplex halimus* L. Australian J. Botany **16**, 469–478 (1968).

FAHN, A.: Plant Anatomy. Second Edition. Oxford: Pergamon Press 1974.

GALE, J., NAAMAN, R., POLJAKOFF-MAYBER, A.: Growth of *Atriplex halimus* L. in sodium-chloride salinated calture solutions as affected by the relative humidity of the air. Aust. J. Biol. Sci. **23**, 947–952 (1970).

GALE, J., POLJAKOFF-MAYBER, A.: Interrelations between growth and photosynthesis of salt bush (*Atriplex halimus* L.) grown in saline media. Australian J. Biol. Sci. **23**, 937–945 (1970).

GOODMAN, P. J.: Physiological and ecotypic adaptation of plants to salt desert conditions in Utah. J. Ecol. **61**, 473–494 (1973).

GOREV, L. N.: Ecologo-physiological basis for vine cultivation in Uzbekistan on soils with high water table (in Russian). Autoreferat doktorskoi dissartatsii Akademiya Nauk, SSSR, Moskva (1954).

HAYWARD, H. E., LONG, E. M.: Anatomical and physiological response of the tomato to varying concentrations of sodium chloride, sodium sulphate and nutrient solution. Botan. Gaz. **102**, 437–462 (1941).

JONES, R., HODGKINSON, K. C.: Root growth of rangeland chenopods. Morphology and production of *Atriplex numularia* and *Atriplex vesicaria*. In: JONES, R. (Ed.): The biology of *Atriplex*. CSIRO, Division of Plant Industrj, Canberra, Australia (1970).

MEIRI, A., KOMBAROFF, J., POLJAKOFF-MAYBER, A.: The response of bean plants to sodium chloride and sodium sulphate salinization. Ann. Botany **35**, 837–847 (1971).

MEIRI, A., POLJAKOFF-MAYBER, A.: The effect of chlorine salinity on growth of bean leaves in thickness and in area. Israel J. Botany **16**, 115–123 (1967).

MEIRI, A., POLJAKOFF-MAYBER, A.: Effect of various salinity regimes on growth, leaf expansion and transpiration rate of bean plants. Soil Sci. **104**, 26–32 (1970).

NIR, I., KLEIN, S., POLJAKOFF-MAYBER, A.: Changes in fine structure of root cells from maize seedlings exposed to water stress. Australian J. Biol. Sci. **23**, 489–491 (1970).

OPPENHEIMER, H. R.: Adaptation to drought: Xerophytism. Plant Water Relationships in Arid and Semi-arid Conditions. UNESCO Publication, Arid Zone Research, **15**, 105–138 (1960).

PETROSYAN, G. P.: Cultivation of vines and other fruit crops under the conditions of carbonate salinization in the Araks valley (in Russian). Fiziologiya ustoichivosti rastenii—Trudy Konferentsii 3–7 marta 1959—Akademiya Nauk, SSSR (1960).

POKROVSKAYA, E. I.: Obmen veshchestv rastenii na zasolennykh pochvakh (Metabolism of plants on saline soils). Avtoreferat Kandidatskoi dissartatsii, Moskva (1954).

POKROVSKAYA, E. I.: Nekotorye dannye ob okislitel'no—vosstanovlitel'nykh protzesakh u galofitov (Some data on oxido-reductive processes in halophytes). In: Pamyat'i "Akademika N.A. Maksimova". Akademiya Nauk, SSSR (1957).

RICHARDS, L. A. (Ed.): Diagnosis and improvement of saline and alkali soils. United States Department of Agriculture, Handbook 60 (1954).

SHAHAM-GOLD, N.: Interaction between gibberelin and NaCl in their effect on growth of *Atriplex halimus*. M. Sc. Thesis (in Hebrew), The Hebrew University of Jerusalem (1963).

SHIMONY, C.: The ultrastructure and function of salt secreting glands in *Tamarix aphylla* L. and *Avicennia marina (Forssk) Vierh.* Ph. D. Thesis (in Hebrew), The Hebrew University of Jerusalem (1972).

STROGONOV, B. P.: Fisiologithcheskie Osnovy Soleustoitchivosti Rastenii (Physiological bases of salt tolerance in plants). Akademia Nauk SSSR, Moskva (1962).

STROGONOV, P., KABANOV, V. V., SHEVJAKOVA, N. I., LAPINA, L. P., KOMIZERKO, E. I., POPOV, B. A., DOSTANOVA, R. KH., PRYKHOD'KO, L. S.: Struktura i Funktziya Kletok rastenii pri Zasolenii (Structure and function of plant cells under salinity). Nauka, Moskva (1970).

UDOVENKO, G. V., GRADCHANINOVA, O. D., SEMUSHINA, L. A.: Morfologicheskye i anatomicheskiye Izmeneniya v listyakh i kornyach pshenizy s uvilicheniyem zosolenya pochvy (Morphological and anatomical changes in wheat leaves and roots with increasing soil salinit). Botanitcheskii J. **55**, 931–937 (1970a).

UDOVENKO, G. V., MASHANSKY, U. F., SINITZKAYA, I. A.: Changes of the root cell ultrastructure in plants with different salt tolerance, during salinization (in Russian). Fisiol. Rastenii **17**, 975–981 (1970b).

WAISEL, Y.: Biology of halophytes. New York: Academic Press 1972.

WEST, K. R.: The anatomy of *Atriplex* leaves. In: JONES, R. (Ed.): The biology of *Atriplex*. Deniliquin N.S.W. Symposium. October 1969. CSIRO, Canberra 11–15 (1970).

WILLERT, D. J., VON, KRAMER, D.: Feinstruktur und Crassulacean-Sauerstoffwechsel in Blättern von *Mesembryanthemum crystallinum* während natürlicher und NaCl-induzierter Alterung. Planta **107**, 227–237 (1972).

WITSCH, H. U., FLÜGEL, A.: Über photoperiodischinduzierte Endomitose bei *Kalanchoe blasfeldiana*. Naturwissenschaften **38**, 138–193 (1951).

CHAPTER VII

The Structure and Function of Salt Glands

W. W. THOMSON

A. Introduction

I. What are Salt Glands and Where Do They Occur?

Many halophytic plants have epidermal glands on their leaves and stems which secrete salt (METCALFE and CHALK, 1950). These glands have been considered efficient devices for the secretion of excess salt which accumulates in the tissue (HABERLANDT, 1914; HELDER, 1956; SCHOLANDER, 1968; SCHOLANDER et al., 1962; 1965; 1966). HELDER (1956) indicated that salt glands were common in the families *Plumbaginaceae* and *Frankeniaceae* but only occurred in a few scattered species outside these families. However, many other plants are known to have trichomes, glands, and glandular structures, but in many instances further investigations are needed to determine their secretion products. Many of these may possibly be salt glands (i.e. specialized structures which secrete minerals and ions) and an understanding of the general distribution and significance of salt glands must await further information.

In surveying the literature it becomes quite apparent that there is a lack of definition concerning salt glands and a persistent confusion between hydathodes and salt glands. In general, hydathodes are recognized as structures of varying degrees of specialization which emit water and solutes to the surface of the plant. However, HABERLANDT (1914) divided hydathodes into two functional types: (a) passive hydathodes—those having a direct communication with the conducting system in which the secretion is a process of filtration under pressure. (b) active hydathodes—those not directly connected with the conducting system which are active in the secretory process. Following the suggestion of STOCKING (1956), in this review I will use HABERLANDT's criteria and identify "active hydathodes" as salt glands with the circumscription, as pointed out above, that they are structures specialized for the secretion of ions and minerals. Thus, for example, no general distinction is drawn between salt glands and chalk glands (METCALFE and CHALK, 1950). It should be pointed out, however, that as suggested by HABERLANDT (1914), detailed comparative studies between hydathodes and salt glands may reveal, transitional forms.

II. What Do Salt Glands Secrete?

The rationale for defining salt glands as those secreting both ions and minerals, thus including chalk glands, is based on two considerations: Firstly, as SCHMIDT-NIELSEN (1965) has pointed out, "... it is generally considered that active transport of water is less feasible than movement of water by osmosis following a primary transport of salt."

Secondly, analyses of the secretion products of salt glands have revealed that a variety of mineral elements are secreted. These include the cations Na, K, Mg, Ca, (SCHTSCHERBACK, 1910; HABERLANDT, 1914; RUHLAND, 1915), and the anions Cl, SO_4, PO_4, and carbonate (VOLKEN, 1884; SCHTSCHERBACK, 1910; HABERLANDT, 1914). The presence of organic material in some cases was indicated by SCHTSCHERBACK (1910) and particularly by RUHLAND (1915), who reported an organic content of 20–40% in the secretion of *Statice*.

Although the early studies indicated that a variety of mineral salts was secreted from salt glands, the general assumption in the literature for a considerable period of time was that the major salt was NaCl (ARISZ et. al., 1955; ARISZ, 1956; HELDER, 1956). Previous assessments of the amounts of substances other than NaCl have been based on a comparison of the osmotic concentration and the chloride content of the secreted fluid. Recent studies have shown that this procedure can lead to serious errors and erroneous assumptions if a rather complete analysis is not done. Analyses of the secretion from the glands of *Tamarix* (WAISEL, 1961; BERRY and THOMSON, 1967; BERRY, 1970), and the mangrove *Aeluropus* (POLLACK and WAISEL, 1970) have shown that the secretion may contain other ions including K, Ca, and Mg. Recently, BERRY (1970) presented a detailed analysis of the crystallized salt from *Tamarix* branches which indicated that Na, K, Mg, and Ca comprised about 99% of the total cations, NO_3, Cl, HCO_3, and SO_4 accounted for about 99% of the anions with bicarbonate being about 60% of the total. Small amounts of micronutrients were also observed. He concluded that since approximately 99% of the salt consisted of the above mentioned ions, very few organic components were secreted from the glands of *Tamarix* (if HCO_3 is considered inorganic). The recurrent suggestion that organic substances are secreted cannot be easily dismissed. Recently SHIMONY and FAHN (1968) concluded from their studies that pectic material was present in the secretion from glands of *Tamarix* and POLLACK and WAISEL (1970) have reported the presence of organic materials, including free amino acids and proteins, in the secretions of *Aeluropus* although they failed to present complete details and results.

B. Structure

I. General Structure

HABERLANDT (1914), in his classical book, pointed out that active hydathodes, or salt glands, may be unicellular as in *Gonocaryum* or multicellular as in *Plumbago* and *Statice*. They may be either sunken in the epidermis, as in the case with *Frankenia* (Fig. 1), or in some instances protrude as hairs and trichomes. RUHLAND (1915) found that in *Statice* the salt gland was composed of a complex of sixteen cells which contain a dense cytoplasm, large nuclei, and many small vacuoles. Interior but adjacent to the gland were four large, highly vacuolated cells which RUHLAND termed collecting cells. Except for the walls between the innermost gland cells and the collecting cells, identified by RUHLAND as a transfusion zone, the gland was encapsulated by a thick cuticular layer containing small pores in its outer mantle. Because of this cuticular encasement, the transfusion zones appeared to be the only area of possible connection between the collecting cells and the cells of the gland.

CAMPBELL and STRONG (1964) reported that the salt glands of *Tamarix pentandra* develop from a single protoderm cell and consist of six secretory cells, with granular cytoplasm and large nuclei, and two inner, vacuolate collecting cells. These observations were supported by SHIMONY and FAHN (1968) who also presented a drawing illustrating the gland encapsulated by a cuticular layer.

The studies of SKELDING and WINTERBOTHAM (1939) indicated that the salt gland in the grass *Spartina* differs considerably from the glands of *Statice* or *Tamarix*. They reported that in *Spartina* the gland consists of only two cells, a small, outer cap cell positioned on a neck-like protrusion of a large basal cell. Both cells contain a granular cytoplasm. The outer shoulders of the basal cell abut on four epidermal cells and the inner walls connect with the adjacent mesophyll cells. The outer boundary of the gland is heavily cuticularized with the cuticular layer continuous with the cuticle of the adjacent epidermal cells. However, the gland is not enclosed by the cuticular layer as occurs in *Statice* and *Tamarix* (LEVERING and THOMSON, 1971). SKELDING and WINTERBOTHAM described the presence of pits in walls between the basal cell and mesophyll cells and in the walls connecting the basal cell with the epidermal cells.

The vesiculate hairs or bladders of *Atriplex* have received considerable attention relative to salt secretion. These trichomes have a large, highly vacuolate bladder cell attached to a stalk composed of one or more cells which in turn is attached to an epidermal cell (BLACK, 1954; OSMOND et al., 1969; MOZAFAR and GOODIN, 1970). In these vesiculated hairs, the emission of salt is apparently the result of the rupturing and collapse of the bladder cells (OSMOND et al., 1969, LÜTTGE, 1971). Although this might not be considered strictly a secretory process, these trichomes are considered salt glands in this review since their function is obviously a specialized mechanism for the removal of salt from the leaves.

II. Ultrastructure

Electron microscopic studies have corroborated many of the light microscopic observations and have provided much new insight and information. A comparison of the ultrastructural features of salt glands studied to date suggests a possible differentiation between three types of glands.

The first type includes glands of *Limonium* (Fig. 6) *Tamarix* (Figs. 2 and 9), *Frankenia* (Fig. 1), and many of the mangroves (Figs. 3–5 and 7). One of the characteristic features of these glands is that they are almost entirely enclosed by a cuticular layer (Figs. 1–7, *cu*) except for small regions, the transfusion zones, between the innermost gland cells and the sub-basal cells. This feature, described in light microscope studies by RUHLAND (1915), was demonstrated in electron microscopic examination of *Limonium* (ZIEGLER and LÜTTGE, 1966 and Fig. 6, *t*), *Tamarix* (THOMSON and LIU, 1967; THOMSON, BERRY and LIU, 1969; SHIMONY and FAHN, 1968, and Figs. 2 and 9, *t*), *Aegialitis* (Fig. 7, *t*, and ATKINSON et al.,

1967), and is illustrated here for *Frankenia* (Fig. 1, *t*), *Avicennia* (Fig. 3, *t*), *Acanthus* (Fig.4, *t*), and *Aegiceras* (Fig.5, *t*). The cuticle is frequently separated from the walls of the secretory cells along the outer surface of the gland (Figs. 1–7, *c*) creating a large, generally electron transparent cavity or collecting compartment between the secretory cells and the cuticle. The inner portion of the cuticle is often loosely organized along the summit of the cuticular cap (Figs. 3 and 4) and electron dense material is frequently observed in the irregular gaps in this region (Fig. 3). RUHLAND (1915) described the presence of pores in the cuticular layer of *Statice* and such pores, often containing dense material, have been observed in electron microscopic studies of *Limonium* (ZIEGLER and LÜTTGE, 1966) and *Tamarix* (THOMSON and LIU, 1967; SHIMONY and FAHN, 1968, and Fig. 2, *p*). The cuticular layer extends over the gland and inward along the lateral sides of the gland where it abuts with the walls of the collecting cells or sub-basal cells. In most cases the walls of the secretory cells are not contiguous with the walls of the collecting and sub-basal cells (Figs. 1–7, *lc*). In many glands where the cell wall is absent along the lateral walls of the innermost gland cells or basal cells, an unusual interfacial structure occurs (Fig. 8). In this region the cuticular material is more electron dense than the associated encapsulating layer and small fibrils are often observed in the cuticular matrix. Although irregular blebs of the plasmalemma are a common feature (Fig. 8, *b*), the plasmalemma has close contact with the cuticular material particularly near the juncture of the basal cells with the adjacent secretory cells. An extremely important point, particularly in regards to interpretations concerning the mechanism of secretion, is that the walls of the collecting cells or sub-basal cells do not appear to be cuticularized at any point in and near the transfusion zone. This is true for *Frankenia*, (Fig. 1, *), *Tamarix* (Fig. 2, *), the mangroves (Figs. 3, 4, and 7, *), and shown in Fig. 6, * for *Limonium*. This is also apparent in the micrographs of *Limonium* published by ZIEGLER and LÜTTGE (1966, 1967).

Characteristically, numerous plasmodesmata occur in the transfusion zone wall between the inner secretory cells and the collecting cells (Fig. 2, *pl*, and ZIEGLER and LÜTTGE, 1966; 1967; THOMSON and LIU, 1967; THOMSON et al., 1969; ATKINSON et al., 1967). The secretory cells of the gland are also interconnected by plasmodesmata.

In some glands numerous wall protuberances occur along the walls of the secretory cells. This is particularly evident in *Frankenia* (Fig. 1, *wp*) and *Tamarix* (Figs. 2, and 9, *wp*, and THOMSON and LIU, 1967; THOMSON et al., 1969; SHIMONY and FAHN, 1968) and *Spartina* (Figs. 13, and 14, *wp*). Wall protuberances also occur in the secretory cells of *Limonium* (ZIEGLER and LÜTTGE, 1966, 1967) although judging from the published micrographs the number appears to be considerably less than in *Tamarix* and *Frankenia*. Occasional small protuberances are observed in *Aegialitis*, but they appear to be absent in the glands of *Avicennia*, (Fig. 3), *Acanthus* (Fig. 4), and *Aegiceras* (Fig. 5). The absence of these structures in some glands seems contrary to LÜTTGE'S (1971) conclusion that cell-wall protuberances are required for intensive short-distance transport. Particularly damaging to this thesis is the infrequency of wall protuberance in the glands of *Aegialitis* (Fig. 7) which apparently has one of the most active salt secretion mechanisms in either animals or plants (ATKINSON et al., 1967).

The protuberances in *Frankenia* (Fig. 1) and *Tamarix* (Fig. 2) are abundant in the outer secretory cells, somewhat reduced in number in the middle secretory cells, and virtually absent from the inner secretory cells (THOMSON and LIU, 1967; SHIMONY and FAHN, 1968). ZIEGLER and LÜTTGE (1966, 1967) have reported that in the salt glands of *Limonium*, some of the wall protuberances were associated with plasmodesmata between the secretory cells but unequivocal electron microscopic evidence was not presented.

As pointed out by several authors (ZIEGLER and LÜTTGE, 1966; THOMSON and LIU, 1967; ATKINSON et al., 1967) and illustrated here (Figs. 1–5, 7, *co*), the collecting cells are ultrastructurally very similar to the underlying mesophyll cells. They have large vacuoles often containing electron dense material, well developed chloroplasts and a peripheral cytoplasm with mitochondria and other organelles. Since the collecting cells or sub-basal cells are almost identical in appearance to mesophyll cells and are strikingly different from the secretory cells, except for location, it would seem they may have only a small role in the secretory process. However, CAMPBELL and STRONG (1964) reported that in *Tamarix* the entire gland, including the collecting cells, develops from a single protoderm cell and ATKINSON et al. (1967) have described in *Aegialitis* a concentration of mitochondria near the plasmodesmata in the wall between the sub-basal cell and the basal cell of the gland. Our studies have not substantiated this observation (Fig. 7) although it is possible that localized accumulations of mitochondria may be missed in thin sections.

The main gland cells are characterized by a dense cytoplasm, many mitochondria, large nuclei and small vacuoles, frequently with an electron dense content (Figs. 1–7, and ZIEGLER and LÜTTGE, 1966, 1967; THOMSON and LIU, 1967; THOMSON et al., 1969; SHIMONY and FAHN, 1968; ATKINSON at al., 1967). Variation in the ultrastructure of different cells within the gland has been reported and variation in the fine structure of the secretory cells exists between the different glands. THOMSON and LIU (1967) reported that in the glands of *Tamarix* the number of mitochondria was greatest in the outer secretory cells, less in the middle cells, and still less in the inner secretory cells. However, SHIMONY and FAHN (1968) reported that the relative number of mitochondria in the outer and middle secretory cells was the same but the inner secretory cells had less. The glands (Fig. 1) of *Frankenia* of the *Frankeniaceae*, a closely related family to *Tamariaceae*, appear similar to those of *Tamarix* in most respects (compare Fig. 1 with Fig. 2) but the secretory

Abbreviations Used on Micrographs (Fig. 1—14)

b — membrane blebs;
c — collecting compartment;
co — collecting cells;
cu — cuticle;
er — endoplasmic reticulum;
ic — inner cuticle
lc — lateral cuticle;
m — mitochondrion;
n — nucleus;
p — pores
pd — plastid;
pl — plasmodesmata;
pm — partitioning membranes;
t — transfusion zone;
v — vacuole;
w — wall;
wp — wall protuberance

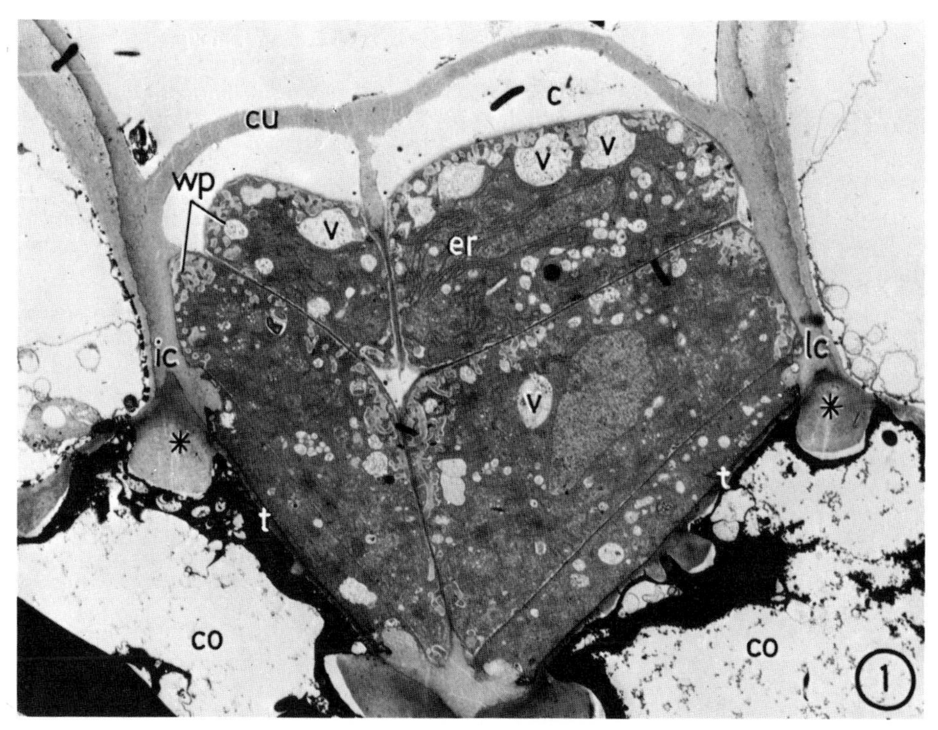

Fig. 1. Salt gland of *Frankenia*. Note the dense cytoplasm, endoplasmic reticulum, *er*, and the large collecting compartment, *c*. ×2900

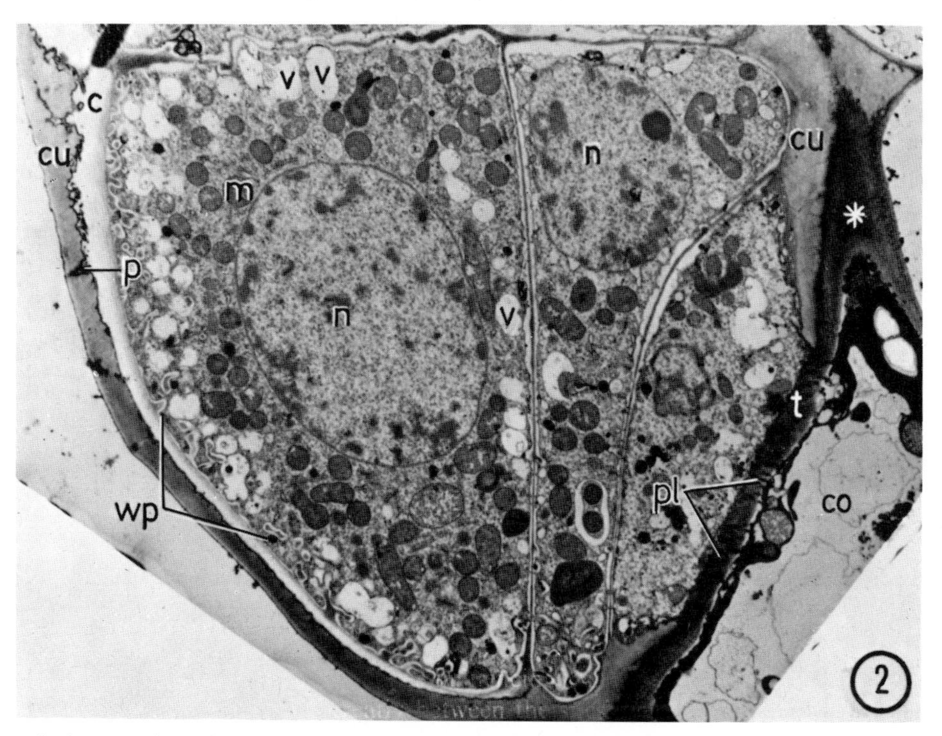

Fig. 2. A portion of a salt gland of *Tamarix*. There are numerous mitochondria, *m*, in the secretory cells and many small vacuoles, *v*, associated with the walls and wall protuberances, *wp*, of the secretory cells. It can be easily seen that the cuticle, *cu*, does not block the wall between the asterisk, *, and the transfusion zone, *t*. ×5800

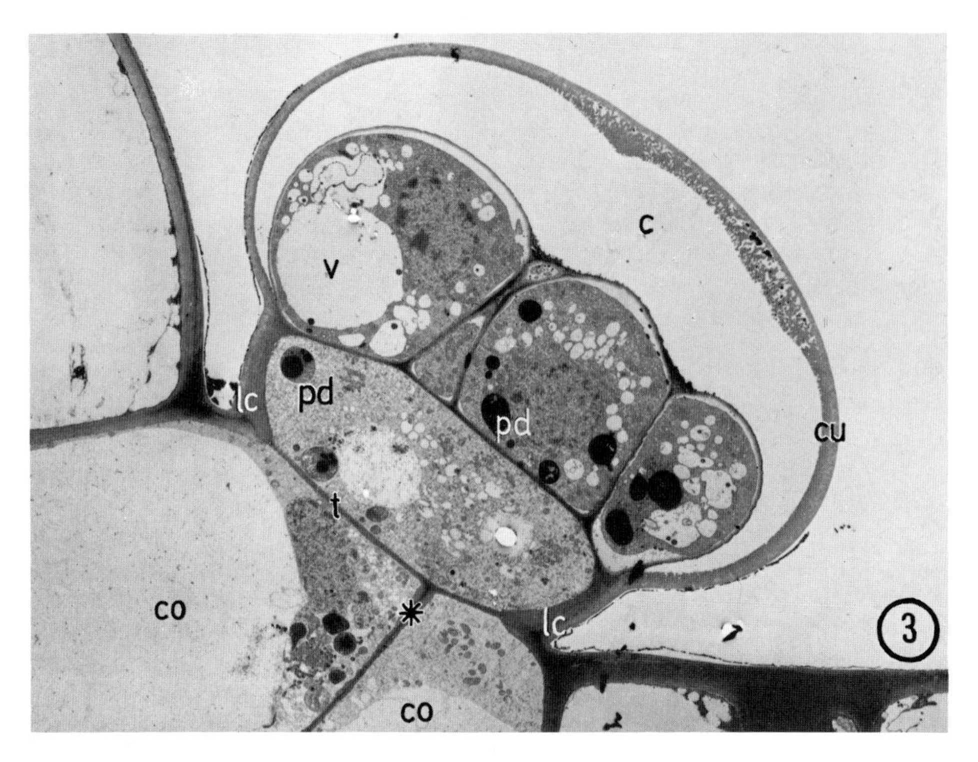

Fig. 3 The salt gland of *Avicennia*. Note the large collecting compartment, *c*, and the dense plastids in the secretory cells. × 1900

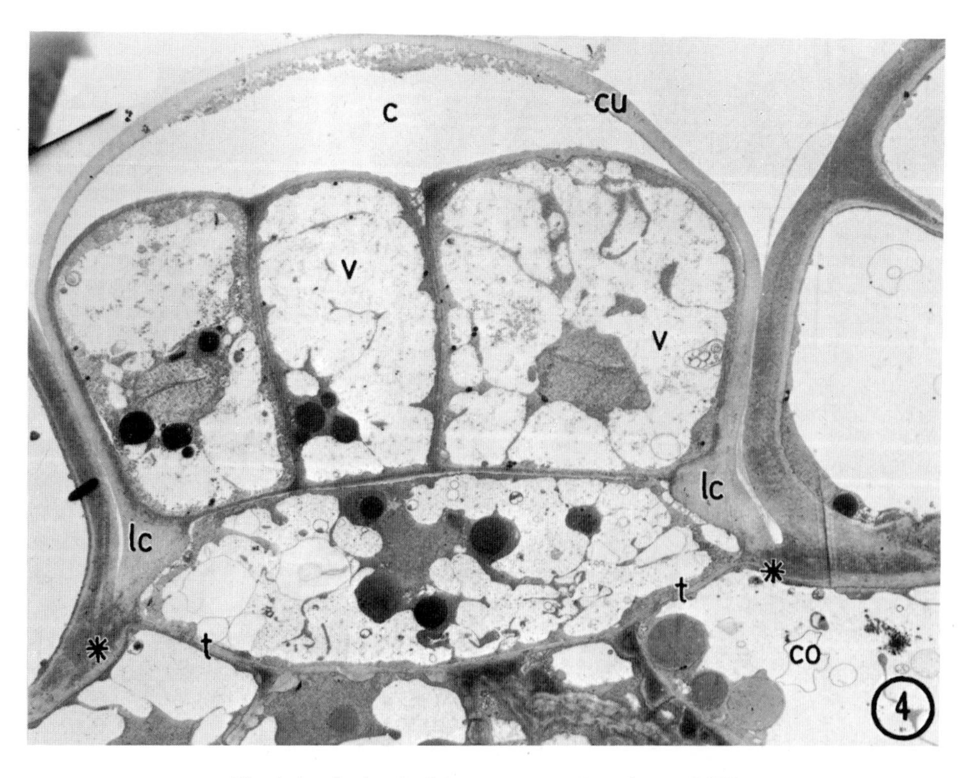

Fig. 4. A salt gland of the mangrove *Acanthus*. × 2650

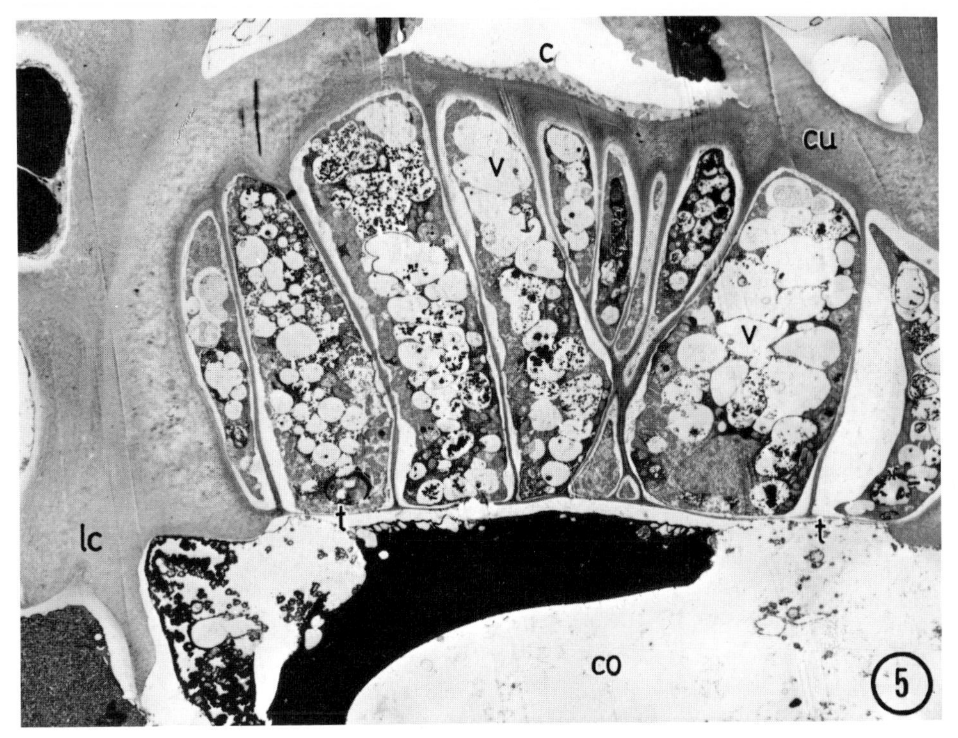

Fig. 5. A part of the salt gland of the mangrove, *Aegiceras*. The secretory cells have numerous vacuoles. × 1800

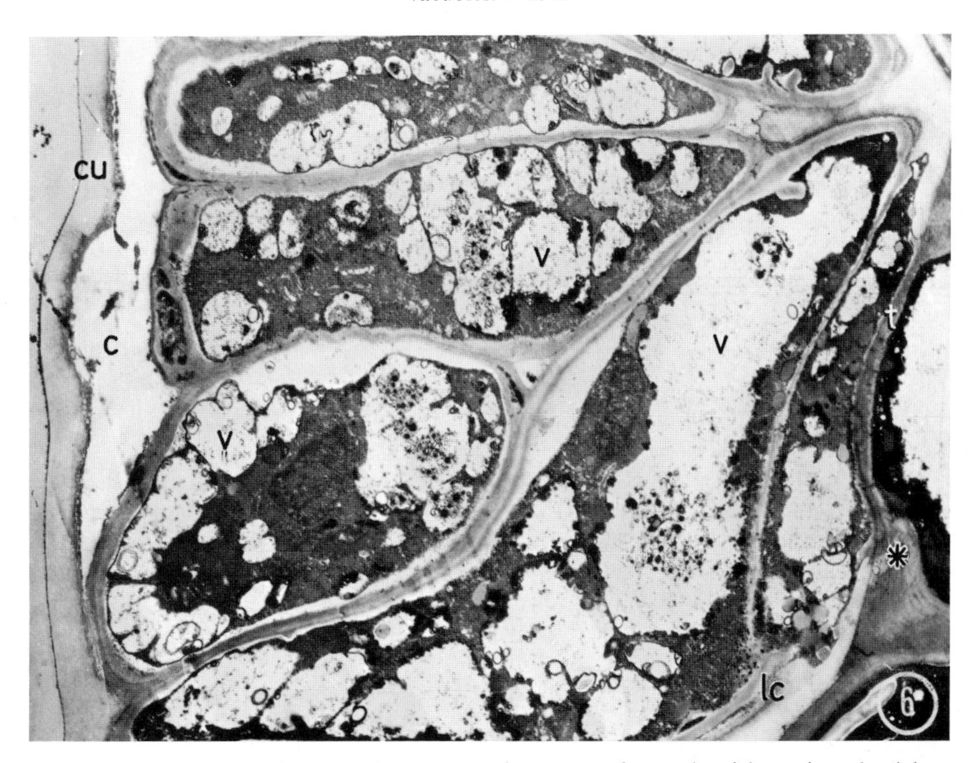

Fig. 6. A portion of the salt gland of *Limonium*. The outer surface and cuticle, *cu*, is to the right, while the inner transfusion zone, *t*, is to the left. Note the wall between the asterisk, *, and the transfusion zone, *t*, is not encrusted by the cuticle, *lc*. × 4800

cells in the glands of *Frankenia* appear to have denser cytoplasm. Furthermore, large accumulations of rough endoplasmic reticulum occur in the secretory cells in the glands of *Frankenia* particularly in the outer secretory cells (Fig. 1, *er*). Such quantities of reticulum were not observed in secretory cells of *Tamarix* (Fig. 1) or in other glands of this type (Figs. 3–7). ATKINSON et al. (1967) reported that the endoplasmic reticulum was abundant and associated with the cell wall of the basal cells in the glands of *Aegialitis*. We have not observed any particular distribution of endoplasmic reticulum in the gland cells of *Aegialitis*, although occasional accumulations of endoplasmic reticulum were observed in the inner cap-cell of the gland. ZIEGLER and LÜTTGE (1966) have described the presence of an apparently membraneous electron dense band of material in the secretory cells of *Limonium*.

Vacuoles have been observed in almost all secretory cells of this type of gland (Figs. 1–7, *v*). The vacuoles often vary considerably in size, even within the same cell, and they range from being electron translucent to being completely filled with an electron dense material (Figs. 1–7, *v*, and ZIEGLER and LÜTTGE, 1966; THOMSON and LIU, 1967; SHIMONY and FAHN, 1968; ATKINSON et al., 1967). In their studies on the ultrastructure of the salt glands of *Tamarix*, THOMSON and LIU (1967) reported that the number of vacuoles in the secretory cells varied. Vacuoles were almost totally absent in some glands, while in others there were numerous vacuoles, many of which were distributed along the periphery of the cells associated with the wall protuberances (Fig. 2, *v*). A similar distribution of vacuoles was reported by SHIMONY and FAHN (1968) in their studies of the glands of *Tamarix*.

Plastids occur in all secretory cells of the glands. They have little internal membrane development consisting mostly of small vesicles and some lamellae particularly in the peripheral regions of the organelles. In *Tamarix* (SHIMONY and FAHN, 1968) and occasionally in other glands, the plastid lamellae are often arranged somewhat concentrically. The plastids generally have a very dense stroma and in the outer secretory cells of *Avicennia* the plastid matrix is often completely opaque (Fig. 3, *pd*). Except in situations where the plastids are almost totally electron dense, plastoglobuli are commonly observed.

The second type of gland to be considered is the trichome of *Atriplex*. OSMOND et al. (1969), from their electron microscopic studies, represent the gland as covered by a cuticular layer with the lateral walls of the stalk cell as completely cuticularized and/or suberized (LÜTTGE, 1971). Whether or not the lateral walls of

Fig. 7. A slightly off median section of the salt gland of *Aegialitis*. There are numerous vacuoles in the secretory cells. × 1900

Fig. 8. A portion of a salt gland of *Tamarix* showing the lateral cuticle, *lc*, and interfacial region between the inner transfusion zone, *t*, the wall, *w*, between the secretory cells of the gland. Note that the cell wall is not continuous from the wall, *w*, of the secretory cells to the transfusion zone, *t*. Also the cuticle, *lc*, does not encroach into and "block" the wall at the transfusion zone. × 26500

Fig. 9. An active salt gland of *Tamarix* from a branch which had been placed in 3% NaCl and salt was accumulating on the surface of the leaves. Note the dense cytoplasm and numerous small vacuoles, *v*, in association with wall protuberances, *wp*. × 4500

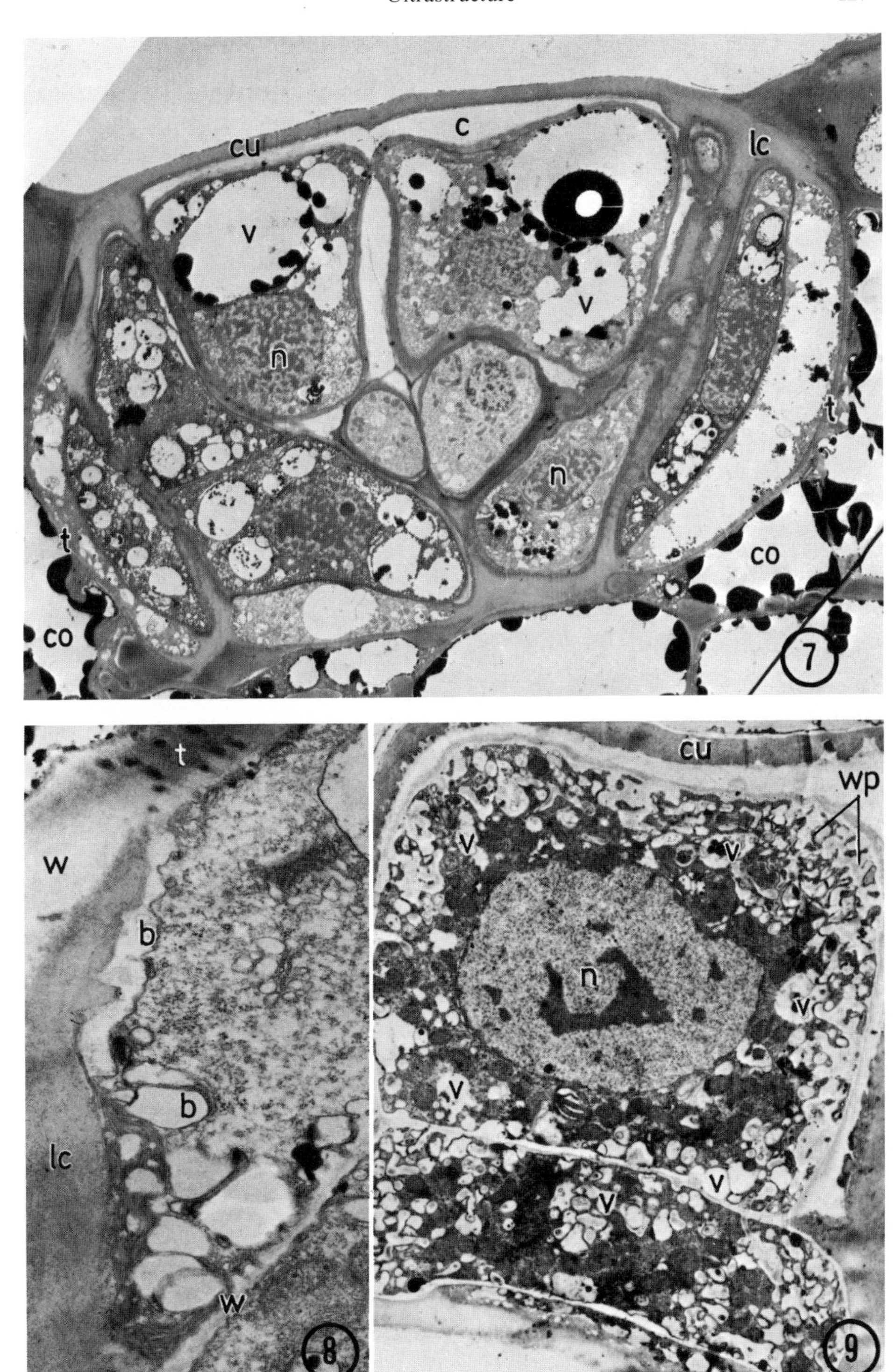
cu
c
lc
v
v
n
n
t
t
co
co
7
t
w
b
b
lc
w
8
cu
wp
v
v
n
v
v
v
v
9

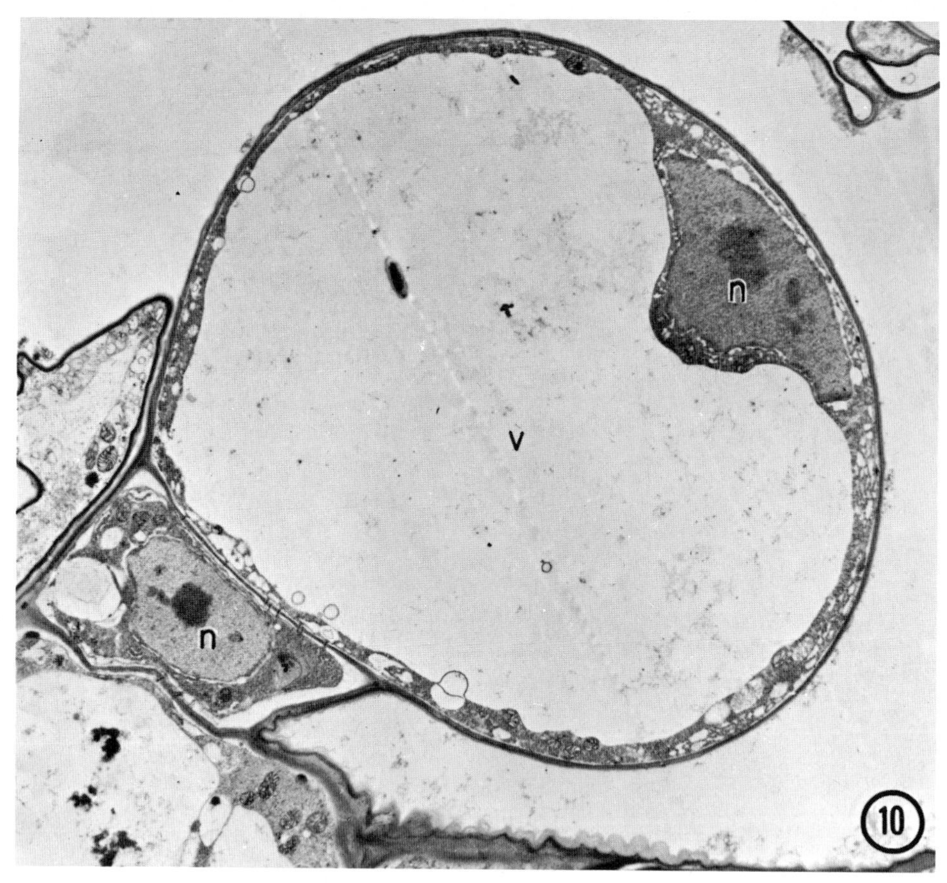

Fig. 10. The stalk and bladder cell of the vesiculated trichome of *Atriplex*. ×6700

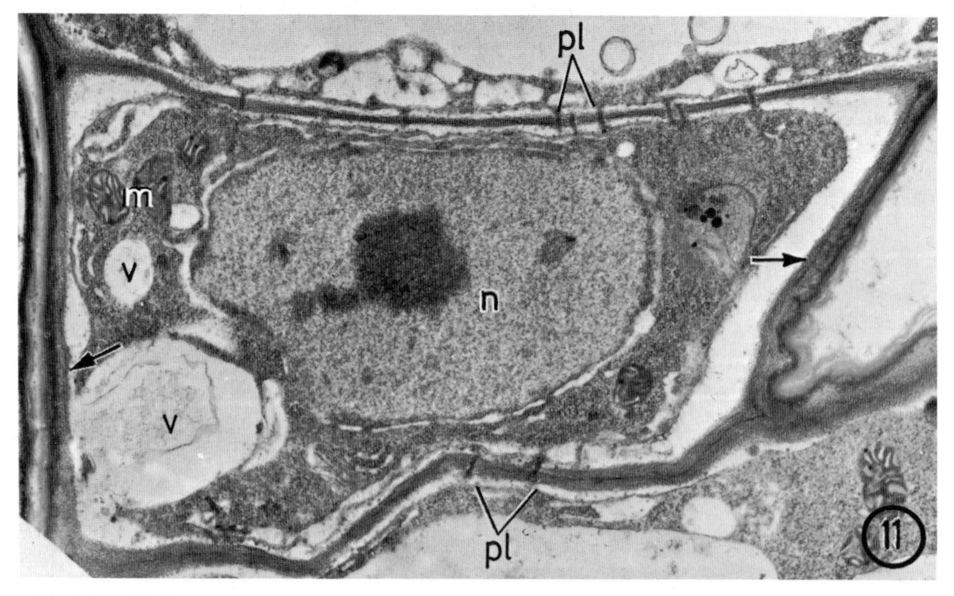

Fig. 11. A higher magnification view of the stalk cell of the *Atriplex* trichome. Of particular interest is the absence of cuticular material along the inner portion of the lateral walls — arrows. × 18 500

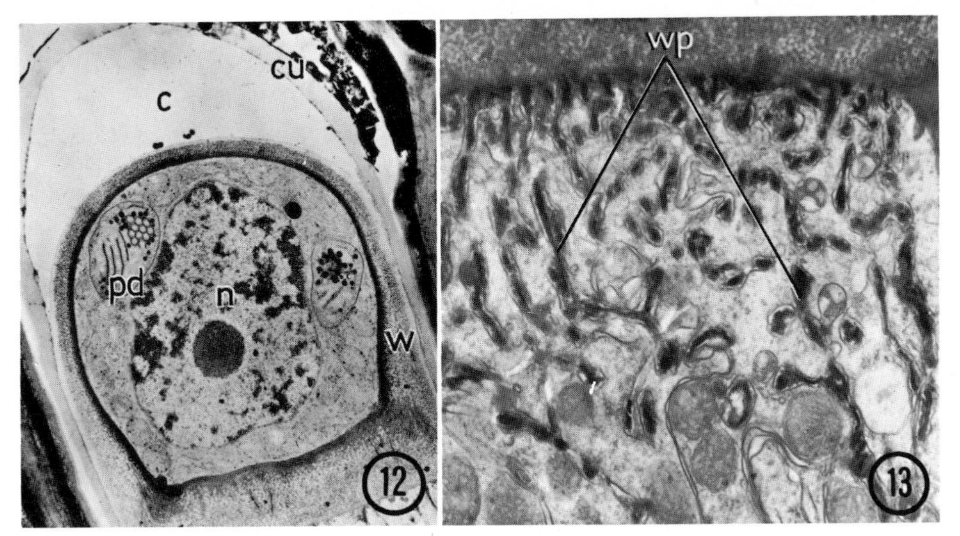

Fig. 12. The cap cell of the gland of *Spartina*. Note the large collecting cavity, *c*. ×6200. By permission of LEVERING and THOMSON, [Planta **97**, 183 (1971)]

Fig. 13. A high magnification of the wall protuberances which extend from the wall between the basal and cap cell into the basal cell. ×15000

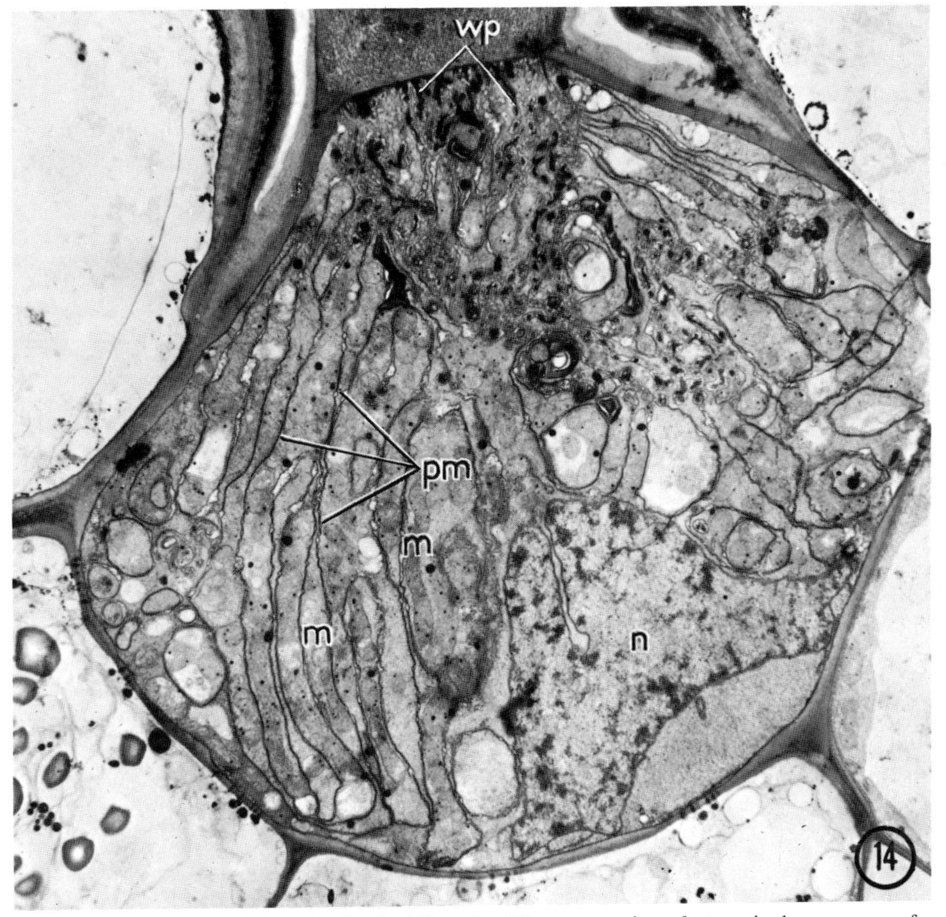

Fig. 14. The basal cell of the salt gland of *Spartina*. The most unique feature is the presence of numerous partitioning membranes, *pm*, associated with mitochondria, *m*. ×6500

the stalk cells are cuticularized is an important point relative to concepts concerning the mechanism of secretion (LÜTTGE, 1971, and following discussion of mechanisms). A careful examination of OSMOND et al.'s published micrographs does not reveal the inner portion of the lateral walls of the stalk cell as being completely cuticularized (e.g. see Figs. 13 and 15 of OSMOND et al., 1969). Recently, electron microscopic studies by SMAOUI (1971) indicate that these walls may not be entirely cuticularized, and that they have an outer electron dense cuticular layer, a middle layer, and an inner striated layer. Our studies on the ultrastructure of the trichomes of *Atriplex*, support the observations of SMAOUI that the inner portion of the stalk cell walls are not cuticularized (Fig. 11, arrow).

OSMOND et al. (1969) and SMAOUI (1971) have reported the presence of plasmodesmata in the wall between the epidermal cell and the stalk cell and in the wall between the stalk cell and the bladder cell (Fig. 11, *pl*). The stalk cell has numerous mitochondria (Fig. 11, *m*), small vacuoles (Fig. 11, *v*), and a dense cytoplasm. OSMOND et al. (1969) state that the stalk cell is packed with endoplasmic reticulum although this is not clearly evident in their micrographs nor have we observed this feature in our studies (Fig. 11). The stalk cells may have chloroplasts with a modestly developed internal membrane system of grana and interconnecting membranes (OSMOND et al., 1969; SMAOUI, 1971).

The mature bladder cell has a large central vacuole (Fig. 10, *v*) with a peripheral cytoplasm containing mitochondria, small vesicles, dictyosomes, and chloroplasts (OSMOND et al., 1969; SMAOUI, 1971).

Because of the unusual ultrastructural features of the two-celled gland of *Spartina*, this gland obviously represents a third structural type. LEVERING and THOMSON (1971) found that the external surface of the gland was covered by a thick cuticle (Fig. 12, *cu*), but the walls of the underlying basal cells were not cuticularized (Fig. 14). The cuticle was expanded outward over the cap cell creating a cavity between the cuticular layer and the cell wall of the cap cell (Fig. 12, *c*). LEVERING and THOMSON (1971) observed thin, electron dense strands across the cuticle suggesting the presence of small pores. The wall of the cap cell was relatively thick and had a "flaky" appearance, being quite dense near the plasmalemma and more diffuse in the outer portions (Fig. 12, *w*). Plasmodesmata occurred in the thick wall between the basal cell and cap cell and between the basal cell and the adjacent mesophyll cells. No plasmodesmata were observed in the wall between the basal cell and the overlying epidermal cells.

Wall protuberances, bordered by the plasmalemma, extended into the basal cell from the wall between the basal and cap cell (Fig. 13 and 14, *wp*). The cytoplasm of the basal cell appears packed with a system of paired membranes, "partitioning membranes", extending throughout the cell (Fig. 14, *pm*). LEVERING and THOMSON (1971) have shown that these paired membranes are extensions of the plasmalemma from the wall protuberances and thus the space between the paired membranes constitutes an extracellular channel or compartment. These channels frequently appeared distended in outline and are associated with numerous mitochondria (Fig. 14).

The nucleus of the basal cell was quite large (Fig. 14, *n*). Plastids containing plastoglobuli, a dense stroma, and a few peripheral vesicles and membranes were also present. Dictyosomes, although not conspicuous, were observed along with free ribosomes and endoplasmic reticulum. The dominant features of the cap cell were a large nucleus (Fig. 12, *n*) and a dense cytoplasm containing mitochondria and plastids with plastoglobuli and a few internal membranes often aggregated into a hexagonal lattice (Fig. 12, *pd*).

C. Function

I. Pathway to and from the Gland

Since the glands are not in direct connection with the vascular tissue, the possible pathway of salt to the gland has been of serious concern. ARISZ (1956) and HELDER (1956) reviewed the structural and physiological studies of salt glands and concluded that due to the enclosure of the glands by a thick, impermeable cuticle (RUHLAND, 1915), apoplastic transport was not possible and only symplastic transport occurred between the collecting cells and the innermost cells of the gland. This concept includes the idea that the cuticular material is functionally and morphologically analogous to the casparian strip in root endodermal cells. ARISZ and colleagues and also HELDER have emphasized that the mechanism of salt secretion and ion transport may be similar to transport across the root. This idea has been further advanced by LÜTTGE (1971) in his review on the mechanism of secretion. However, ultrastructural studies indicate that apoplastic transport to the glands cannot be ruled out. For example, in *Spartina* the inner walls of the basal cell are free from cuticular material and are contiguous with the walls of the underlying basal cells; in *Atriplex*, the lateral walls of the stalk cells and the wall of the bladder cell do not appear entirely cuticularized (Fig. 11 and SMAOUI, 1971); in the *Limonium-Tamarix* type of glands although the lateral walls of the basal cells of the gland are cuticularized, no cuticular block is apparent in the wall between the transfusion zone and the walls of the adjacent mesophyll cells. Consequently, on anatomical grounds, as THOMSON and LIU (1967) have pointed out for *Tamarix*, apoplastic transport to the gland is a distinct possibility.

The occurrence of numerous plasmodesmata between the collecting cells, and the inner or basal cells of the gland, as seen in *Tamarix*, (Fig. 2, *pl*), coupled with the assumption that apoplastic transport could not occur, has led to the suggestion that symplastic transport of ions through the protoplasmic channels of the plasmodesmata is of major importance (ZIEGLER and LÜTTGE, 1966, 1967; LÜTTGE, 1969, 1971; LARKUM and HILL, 1970). Another possible role for the plasmodesmata are passageways for the movement of metabolic substrates to the glands. As suggested by the high complement of mitochondria, the gland cells are undoubtedly highly active and since they lack well developed chloroplasts, except possibly in *Atriplex* (OSMOND et al., 1969), they must be dependent on adjacent tissues for metabolic substrates.

Some support for the concept that the plasmodesmata are involved in the passage of ions to the gland has come from studies of ZIEGLER and LÜTTGE (1967) on the electron microscopic localization of chloride in leaves of *Limonium*. They used the technique devised by KOMNICK (1962), which involves the use of silver acetate in conjunction with osmium tetroxide or potassium permanganate in fixing material for electron microscopy and observed fine precipitates of AgCl along the plasmalemma, along the endoplasmic reticulum and in the nucleus and mitochondria. There were also large precipitates located in the small vacuoles of the gland cells. They reported the presence of precipitates in the plasmodesmata between the inner gland cell and the collecting cell, although close examination of their published micrograph does not clearly illustrate this. Since the apoplastic route to the gland is not blocked by cuticular material, as has been assumed, and the certainty of the localization of chloride in the plasmodesmata is still unresolved, the question of how ions move to the glands, whether through the apoplast, symplast, or both, is still not answered.

The movement of salt and ions out of the glands to the surface of the leaves has also been considered. RUHLAND (1915) observed pores in the outer cuticle covering the glands of *Statice*. More recently in electron microscopic examinations, structures resembling pores have been observed in *Statice* (ZIEGLER and LÜTTGE, 1966, 1967), *Tamarix* (THOMSON and LIU, 1967; SHIMONY and FAHN, 1968), and *Spartina* (LEVERING and THOMSON, 1971). In their electron microscopic study of the localization of chloride, ZIEGLER and LÜTTGE (1967) observed heavy precipitates of AgCl in the pore regions both under and on the outer surface of the cuticle. These observations seem to leave little doubt that the passage of salts out of the glands, particularly of the *Limonium-Tamarix* and *Spartina* types, is via cuticular pores.

Another question in relation to the passage of salts out of the gland concerns the often observed large cavity between the cuticle and the underlying secretory cells (Figs. 1–7 and 12, *c*). Similar cavities have been observed in other types of glandular trichomes and hairs (ESAU, 1965). In recent publications on nectar production in *Abutilon* hair cells, FINDLAY and MERCER (1971 a, b), have elegantly shown that as exudation proceeds, the cuticle expands above the terminal cell, forming a cavity in which the nectar solution apparently collects. Subsequently, small droplets of nectar are rapidly emitted through the cuticle and the expanded cavity collapses. FINDLEY and MERCER (1971 a) concluded that the secretions collect in the subcuticular cavity and when a sufficient hydrostatic pressure develops, the fluid is released through pores which are valve-like and possibly pressure sensitive. The similarity of the cuticular cavity in salt glands to that of *Abutilon* suggests it is also a collecting compartment. Whether the salt solution is released by a mechanism similar to that proposed by FINLEY and MERCER (1971 a) for *Abutilon* is not known but it seems likely, particularly in considering RUHLAND'S (1915) early evidence that secretion occurs under pressure.

Most hypotheses to date recognize that once the salts are released into the walls of the secretory cells and the subcuticular cavity, backflow into the mesophyll tissue is blocked (see LÜTTGE, 1971 for literature review). This conclusion is based on the presence of a cuticular band along the wall surface of the basal or innermost secretory cells, particularly in the *Limonium-Tamarix* type glands (Figs. 1, 2, and 4–7, *l c*). Since in these glands this region appears to be completely cuticularized, and because of the presumed impermeability of this region to aqueous solutions, no direct passageway appears to be open for the movement of salts back into the underlying tissues. Thus, the cuticularized zone apparently has a role relative to ion transport similar to that proposed for the Casparian strip in the endodermis of roots (ARISZ, 1956; LÜTTGE, 1969, 1971). Although direct evidence that this region is an effective barrier has not been presented, its similarity to and continuity with the cuticle of the adjacent epidermal cells, and the morphological observations of the lack of direct connections between walls of the secretory cells and the walls of the underlying collecting cells, provide substantial support for this hypothesis. However, studies are needed to determine whether the plasmalemma associated with the cuticle in this region represents a tight junction, as has been reported for the root endodermis (BONNETT, 1968).

Although OSMOND et al. (1969) have reported that the lateral walls of the stalk cell in *Atriplex* are cutinized and LÜTTGE (1971) has diagramatically and physiologically represented these walls as comparable to the Casparian Strip, the recent electron microscope studies of SMAOUI (1971) and our studies (Fig. 11, arrow) indicate that these walls are not entirely cutinized. In consideration of the mode of secretion in *Atriplex* SMAOUI's observations are not surprising. In *Atriplex*, the salts apparently accumulate in the large vacuole of the bladder cell (OSMOND et al., 1969; LÜTTGE and OSMOND, 1970; LÜTTGE, 1971; MOZAFAR and GOODIN, 1970) and, with subsequent rupture of the bladder cell, are relased to the surface of the leaf. By this mode of secretion, the salts are not released directly to the walls as is apparently the case for *Limonium-Tamarix* type glands; and the problem of direct backflow into the tissue does not exist.

II. Selectivity and the Ionic and Osmotic Concentration of the Secreted Fluid

The secretion from some salt glands can be quite extensive (RUHLAND, 1915). As much as 70 mg of fluid is secreted from a 150 mg *Limonium* leaf disc in 24 hrs (ARISZ et al., 1955). Even greater rates of secretion, of 90 peq cm^{-2} sec^{-1}, are reported by ATKINSON et al. (1967) for *Aegialitis*. This latter rate is one of the highest known to occur in biological systems (JENNINGS, 1968).

One of the most interesting observations from recent studies is that not only is the secretion from some salt glands composed of different ions, but also that the percentage of the ions varies with the composition of the environment of the roots (WAISEL, 1961; BERRY and THOMSON, 1967; POLLACK and WAISEL, 1970). For example, upon increasing the sodium content of the culture solution, there was an increase in sodium with a concomitant decrease in potassium in the secretion fluid of *Tamarix* (BERRY and THOMSON, 1967; THOMSON et al., 1969). Further, when the ions Rb and Cs, not usually present in the root environment, were added to the culture solution they were secreted from the glands of *Tamarix* (THOMSON et al., 1969) and *Limonium* (HILL, 1967b). BERRY and THOMSON (1967) and THOMSON et al. (1969) have suggested from these studies that the glands were not selective and the cation composition of the secretion was dependent on the composition of the root environment. In more recent studies, BERRY (1970) has again observed that the cation composition of the secretion closely parallels that of the culture medium and further, that the anion composition influenced the relative amounts of cations secreted. Sodium decreased and divalent ions increased in the secretion when the treatment varied from chloride→sulfate→nitrate.

In studies with *Tamarix*, WAISEL (1961) has also observed variation in the ionic composition of the secretion and that amounts of Na, K, and Ca varied with increases or decreases of these ions in the culture solution. POLLACK and WAISEL (1970) observed in *Aeluropus* that secretion of sodium was partially inhibited by potassium, and increasing the sodium content of the culture medium strongly inhibited potassium secretion. Calcium ions, at a concentration of 0.05 M in the culture solution, tended to decrease sodium secretion and stimulate potassium secretion. This stimulation was apparently counterbalanced by sodium when the culture solution contained both sodium and calcium ions. They reported that calcium ions were low in the secretions in all experiments. However, since they performed their analysis only on secreted droplets or from distilled water washing of the leaves, insoluble products such as calcium carbonate probably were not present in the fluids they examined. From their observations, POLLACK and WAISEL (1970) noted that the preference of secretion in *Aeluropus* was Na > K > Ca. Earlier, WAISEL (1961) had observed a sequence of Na > Ca > K for *Tamarix*. These series must however be considered tentative until a more thorough examination of the secretion, particularly of insoluble products, is made and the questions concerning the extent of root selectivity are answered.

III. Active or Passive?

Several types of direct and indirect evidence indicate that an active process is involved in secretion. From an ultrastructural point of view, the high complement of organelles, particularly mitochondria, tends to support this and further suggests that the active step(s) is directly associated with the glands. In early studies, SCHTSCHERBACK (1910) observed that salt was secreted from excised leaves when the petioles were inserted in various salt solutions and that secretion occurred from leaf segments floated on water. These studies were extended by RUHLAND (1915) who observed that the concentration of salt in the secretion varied, depending on the concentration of the medium on which the segments were floated. Similar results were observed more recently by ARISZ et al. (1955). They also reported that when the osmotic concentration of the external solution increased, the amount of secreted fluid decreased, but the osmotic concentration of the fluid increased. Studies on the osmotic concentration of secreted fluid revealed it to be higher than the value for the leaf sap.

Continuing in this line, almost all studies to date have established that the concentration of the secreted fluid is higher than that of the root medium (BERRY, 1970; POLLACK and WAISEL, 1970; SCHOLANDER et al., 1962). In fact BERRY (1970) has reported a 50-fold increase in the concentration of secretion over that of the root medium for *Tamarix*. The salt concentration in the bladders of the trichomes of *Atriplex* was observed by MOZAFAR and GOODIN (1970) to increase with increasing concentrations of the culture solution. In mangroves SCHOLANDER et al. (1962) reported that the concentration of the secretion was 10–20 times that of xylem sap. However, the xylem sap in mangroves probably varies greatly in different mangroves since ATKINSON et al. (1967) have reported that the osmotic pressure of the secreted fluid of the mangrove *Aegialitis* was approximately equivalent to that of the xylem sap. As noted above, ARISZ et al. (1955), using leaf discs of *Limonium*, found that Cl concentration of the secreted fluid was higher than that of the leaf when the discs were floated on a salt solution, and the concentration of the secreted fluid was always higher than the expressed sap of the leaf. Studies of POLLACK and WAISEL (1970) on *Aeluropus* also indicated that the concentration of the secreted fluid was equal to or greater than that of the leaf sap. In *Atriplex*, high concentration of ions also occurs in the bladder cells of trichomes, and OSMOND et al. (1969) determined that the concentration of sodium and chloride was always greater than that of the bathing solution. MOZAFAR and GOODIN (1970) measured the salt concentration in isolated hairs and found that it was about 60 times higher than in the leaf sap. Further, OSMOND et al. (1969) using electrical potential measurements, observed a net flux into the bladders in the light against an electrical gradient.

More recently, HILL and his associates (HILL, 1967a, b; HILL, 1970a, b; LARKUM and HILL, 1970; SHACHAR-HILL and HILL (1970) have shown that when *Limonium* leaf discs, with the lower cuticle removed by abrasion, are floated on salt solution, the gland can secrete Na, K, Cs, Rb, Cl, Br, and I. Using electrochemical techniques HILL has also determined that the ions are actively transported from the glands (HILL, 1967a).

ARISZ et al. (1955) reported that secretion was sensitive to temperature and metabolic inhibitors. Raising the temperature from 5° to 25° C increased the rate of secretion, but the concentration of the salt in the secreted fluid remained about the same. Secretion stopped when oxygen was absent and potassium cyanide and other metabolic inhibitors drastically reduced secretion. ATKINSON et al. (1967) also found that the "uncoupling" agent carbonylcyanide β-chlorophenylhydrazone, applied either to the cut petiole or the surface of the leaves of mangroves, inhibited Cl secretion. However, they have pointed out that the studies provide no information as to whether the effect was directly on the bridging of ion transport with electron transport or indirectly due to a general reduction in the production of essential metabolites.

Briefly summarizing, the large number of mitochondria in the glands, the temperature sensitivity of secretion, the observations on the concentrations of the secreted fluid compared to the leaf, xylem sap and the root medium, the electrochemical studies and the effect of metabolic inhibitors, substantiate the statement that the secretion process is active and at some point there is an input of energy.

IV. Physiological Role

The early suggestion by SCHTSCHERBACK (1910), HABERLANDT (1914), and RUHLAND (1915) that the salt glands function in regulating the salt content of the leaves by secreting excess salt from the tissue, is well documented by the studies in recent years. A major contribution to the confirmation of this concept were the studies of ARISZ and associates (1955). They found that when leaf discs were floated on a salt solution, the Cl concentration of the leaf sap and decreased; when the discs were floated on water the Cl content of the leaf sap was even more reduced. Corroborative evidence has been provided by a number of investigators. SCHOLANDER et al. (1962) found that the salt-secreting mangroves contain significant, but small, amounts of salt in the xylem. Subsequently, SCHOLANDER et al.

(1966) concluded that the salt glands remove excess salt which accumulates in the leaves due to transpiration. In their studies on mangroves, ATKINSON et al. (1967) found that the amount of chloride arriving in the leaves with the xylem sap was balanced by secretion and that the chloride content within the leaves remained relatively constant throughout the day. Similarly, POLLACK and WAISEL (1970) concluded from their studies on *Aeluropus* that almost all sodium ions delivered to the leaves were secreted and there was a low rate of accumulation in the leaf. Further support is also contained in the studies of MOZAFAR and GOODIN (1970) on *Atriplex;* they found that with increasing concentrations of salts in the culture solution, the concentration of salts in hairs increased, but the concentration of the expressed sap of the leaves did not change appreciably.

V. Light Dependency?

Several studies have indicated that salt secretion follows a diurnal pattern of high activity during the day and low activity at night (SCHOLANDER et al., 1962; ATKINSON et al., 1967). However, there is little evidence, for most plants, that the secretion is directly related to photosynthesis. This is supported by the studies of ARISZ et al. (1955) who found that salt secretion could take place from leaf discs of *Limonium* floated on salt solutions in the dark and by the observations of ATKINSON et al. (1967) that salt secretion occurred from *Aegialitis* leaves in the dark. In the studies of ARISZ et al. (1955) it was found that the rate of secretion from leaf discs floated on a solution was stimulated by light. In these experiments with leaf discs floated on a salt solution, the direct relationship to transpiration would appear to be reduced. Whether the stimulation is directly related to a photosynthetic mechanism or to the production of metabolites and metabolic substrates is not known. Interestingly enough, in *Atriplex*, OSMOND et al. (1969) and LÜTTGE and OSMOND (1970) have reported an increase in the chloride concentration of the bladder cells of the trichomes, which is apparently more directly linked to photosynthesis. Using radioactive Cl, OSMOND et al. (1969) found a light-stimulated increase in the chloride content of the bladders which took place against an electrochemical gradient. The time course studies indicated that there was a lapse of several hours prior to the appearance of the light-stimulated uptake (OSMOND et al., 1969). Subsequently, LÜTTGE and OSMOND (1970) and LÜTTGE and PALLAGHY (1969) reported that the light-stimulated chloride transport was inhibited by the Photosystem II inhibitor DCMU and that light-dependent CO_2 fixation was inhibited by DCMU in both the leaf and the bladders. Their studies also indicated that the photosynthetic CO_2 fixation in isolated bladders was much less than in the lamina of the leaf even though the trichome cells have rather well developed chloroplasts.

In further studies it was found that light induced transient membrane potentials in the mesophyll cells and similar changes in the bladders of the gland (LÜTTGE and PALLAGHY, 1969). However, the transient changes were not observed in the isolated epidermis of *Chenopodium*, which, according to LÜTTGE (1971) and LÜTTGE and PALLAGHY (1969) has a similar leaf anatomy to *Atriplex*. Experiments with different wavelengths of light indicate that the transients are related to Photosystem II activity in the chloroplasts (LÜTTGE and PALLAGHY, 1969). From these experimental results, the interpretation has been advanced (OSMOND et al., 1969; LÜTTGE and OSMOND, 1970; LÜTTGE and PALLAGHY, 1969; LÜTTGE, 1971) that the energy for the light-dependent transport into the bladders is linked to photosynthesis in the mesophyll, primarily via Photosystem II. Several factors need careful examination before this hypothesis can be fully accepted. Two of these are: (1) Although there are similar transient changes in the membrane potential in the mesophyll and bladder it is not clearly established whether these are in series or in parallel. (2) Even though the light-dependent CO_2 fixation in the trichomes is low compared to the mesophyll, and this correlates somewhat with the results with DCMU, it does not definitely establish whether the chloroplasts in the cytoplasm of the stalk and bladder cells are directly involved. Even the absence of light-dependent transients in membrane potential in the isolated epidermis cannot be completely accepted as establishing that the transport is dependent on mesophyll photosynthesis. This must await a more complete assessment of whether this lack of a transient membrane potential in the isolated epidermis is related to damage in the stalk and bladder cells. The coincidence between transient changes in the mesophyll and bladders, although suggestive, does not establish that there is an absence of a membrane between them, as suggested by these workers (OSMOND et al., 1969; LÜTTGE and OSMOND, 1970; LÜTTGE and PALLAGHY, 1969; LÜTTGE, 1969, 1971). This evidence must be considered as circumstantial until a direct relationship is shown and until there is a demonstration that the events are in series. Further, the consideration that a contact between the mesophyll and the bladders is only through the symplast is based on the conclusion that apoplastic contact cannot occur because the walls of the stalk cells are completely cutinized. However, as was discussed in the section on ultrastructure, the evidence for this is far from conclusive.

VI. Lag Period

In almost all studies to date when non-pretreated plants, cuttings, or leaf discs are challenged with a salt solution a definite lag period occurs before secretion is observed (ARISZ et al., 1955; POLLACK and WAISEL, 1970; HILL, 1970a, b; SHACHAR-HILL and HILL, 1970; OSMOND et al., 1969; LÜTTGE and OSMOND, 1970). ARISZ et al. (1955) observed that the volume of secretion was apparently independent of the salt level of the leaf and POLLACK and WAISEL (1970) observed with *Aeluropus* that at a concentration of 0.3 M NaCl in the treatment solution, the concentration of the secretion increased but there was no increase in the volume. These workers have suggested that two mechanisms are involved in secretion—one affecting the ionic concentration of the fluid and the second the secretion process itself. The secretion process, they suggest, is activated by a sufficient concentration of the external solution but is independent of its composition.

From a series of studies on *Limonium*, Hill and his colleagues concluded that the lag period was not related to a diffusion resistance or an in-series loading of compartments, but more probably was dependent on the induction of ion pumps.

By electrochemical methods and analysis of fluxes of radioactive ions, HILL (1970a, b) observed that the transit half-time for the secretion of an ion correlates with the half-time for the filling of a tissue compartment. HILL suggests that the tissue compartment is the cytoplasm and that the glands directly regulate the ionic concentration of this compartment. HILL observed a similar lag period when the tissue was challenged by external solutions of different concentrations. Since the lag period was long and because there was no change in the filling rate of the cytoplasm with increasing concentration, HILL has suggested that the transport function was metabolically induced. In recent studies, SHACHAR-HILL and HILL (1970) have presented some experimental evidence to support this hypothesis. Pre-treatment of non-salt-challenged tissue with inhibitors of RNA and protein synthesis inhibited secretion. However, with the exception of cycloheximide, application of these inhibitors to actively secreting glands had no effect. Cycloheximide completely inhibited secretion, which they attributed to possible other effects on the tissue. Although SHACHAR-HILL and HILL (1970) observed no difference in ultrastructure between secreting and non-secreting glands, both of which contained numerous ribosomes and polyribosomes, they concluded from their experiments that the process was inductive.

It is interesting, however, that SHACHAR-HILL and HILL (1970) observed that the sigmoid rise in secretion, after the lag period, had a high temperature sensitivity compared to the lag period. They interpreted these results as indicating that the lag period and the sigmoid rise were two separate physiological activities. If the collecting compartment between the gland cells and the cuticle is a site of ion accumulation and the loading of this compartment is energy dependent (a pump?) a high degree of temperature sensitivity would be expected. This could offer an explanation for the inhibition by low temperature during the sigmoid phase observed by SHACHAR-HILL and HILL (1970).

VII. Mechanism(s)

RUHLAND (1915) found with *Statice*, that when a counter force was applied against the surface of the glands, secretion still occurred under pressure. His studies and those of ARISZ et al. (1955) indicated that the concentration of the secreted fluid was higher than that of the leaf sap and that osmotic work is performed by the glands in the secretion process. ARISZ (1956) and HELDER (1956) both suggested that the secretion was forced out by pressure within the glands. According to this concept the secretion itself is strictly a physical process resulting from an increase in hydrostatic pressure in the gland due to active accumulation of solutes and a resultant flow of water into the cells. The concept that the secretion process is the result of increased hydrostatic pressure and the active process is the uptake of solute into the gland explains, according to ARISZ (1956), ARISZ et al. (1955) and HELDER (1956) the observation that when leaf discs are floated on solutions of high concentration, the amount of secretion is reduced but the concentration of the fluid increases.

According to the hypothesis of ARISZ (1956), the secretion itself is due to the build-up of hydrostatic pressure in the gland, but since the glands and underlying cells are interconnected by plasmodesmata, THOMSON and LIU (1967) have pointed out that a mechanism to prevent backflow must exist. Having observed the presence of numerous microvacuoles in the active glands (Fig. 9, *v*) and reduction or almost total absence of these in glands of *Tamarix* which are non-secreting, they proposed that the salts are sequestered within the vacuoles. Subsequently THOMSON et al. (1969) observed that in Rubidium-treated plants, electron dense accumulations occurred in these microvacuoles; whereas, in sodium or potassium treated plants the microvacuoles were electron translucent or contained only small amounts of electron dense material. These comparative studies tended to support their hypothesis. They also noted that the microvacuoles were frequently associated with the wall protuberances and fused with the plasmalemma of the secretory cells. They suggested that the ions were accumulated in the microvacuoles which then migrate to, and fuse with, the plasmalemma, releasing the ionic contents to the walls. Once released to the walls the salt is effectively removed from the plant at this point since backflow along the walls into the leaf is blocked by the cuticularized zone around the inner secretory cell. The evidence that the electron dense material in the microvacuoles is indeed rubidium has been legitimately criticized by LÜTTGE (1971), and more evidence, probably through electron diffraction or electron probe analysis, is needed to identify the microvacuolar content. Further, LÜTTGE has questioned whether the loading of the microvacuoles and the wall may not be occurring in parallel rather than in series. However, the presence of numerous microvacuoles in the active glands (Fig. 9, *v*) and their absence in non-active glands, the comparative studies with rubidium as an electron dense marker for salt accumulation, and the fusion of the microvacuoles with the plasmalemma of the secretory cells, currently favor the hypothesis over the objections.

The question of whether salts are sequestered in microvacuoles in other salt glands, and whether secretion occurs by a process similar to exocytosis, is at present unanswered. As pointed out in the section on ultrastructure, vacuoles of various size and content have frequently been reported in the secretory cells of the *Limonium-Tamarix*-type glands (ZIEGLER and LÜTTGE, 1966, 1967; ATKINSON et al., 1967; SHIMONY and FAHN, 1968) and in fact several of these investigators have raised the possibility that the vacuoles might be directly involved in a general mechanism of secretion, either as compartments to prevent backflow (ATKINSON et al., 1967) or in the transport phenomenon itself.

In *Atriplex*, the salt is not actually secreted to the surface of the leaf but becomes deposited when the bladder cells of the trichomes rupture (OSMOND et al., 1969; MOZAFAR and GOODIN, 1970; LÜTTGE, 1971). It is particularly interesting to note that the salts apparently are accumulated in high concentration in the vacuole of the bladder cells. It would appear that accumulation in microvacuoles in the *Limonium-Tamarix* type of glands and accumulations in the vacuole of the bladder cells of *Atriplex* are related phenomena (see LÜTTGE, 1971 for discussion). It is not surprising, in a broad sense, that vacuoles may have an important function in the secretory process; particularly when considering the fact that in many plant cells the vacuole is a site of ion accumulation. Further, the possibility that microvacuoles may be involved in the ion accumulation and secretory process itself is strengthened by the recent studies and concepts, developed by MAC-

ROBBIE and others, which show that small compartments may be involved in ion transport (see MACROBBIE, 1971; LÜTTGE, 1971, for reviews and bibliography).

The cuticularized walls of the inner cells of the gland have been considered to be comparable, at least physiologically, to the Casparian strip in root cells. This is one of the basic tenets upon which the concept was developed that transport in the root and salt glands are essentially the same process (ARISZ et al., 1955; ARISZ, 1956; HELDER, 1956; LÜTTGE, 1971). Although this is an attractive hypothesis, there is no physiological evidence that apoplastic transport, at least to the gland, does not occur nor is there any electron microscopic evidence that the walls leading to the glands are cuticularized and therefore inhibitory to apoplastic transport to the transfusion zone between the inner gland cells and the collecting cells. This is also true for the stalk cells of *Atriplex* (see discussion section on ultrastructure and also SMAOUI, 1971). Preliminary experiments in this laboratory (CAMPBELL and THOMSON, unpublished observations) with lanthanum fed to *Tamarix* shoots through the cut end of small stems, indicate that this ion, which cannot penetrate the plasmalemma of cells, accumulates in high concentration in the wall of the transfusion zone between the collecting cells and the innermost gland cells. This clearly indicates that apoplastic transport occurs along the contiguous walls to the transfusion zone and consequently strongly suggests that apoplastic transport to the gland may well be the primary pathway.

As pointed out previously, the cuticularized band around the inner secretory cells probably furnishes a resistance to backflow into the underlying mesophyll. A resistance at this point would be particularly important if the collecting compartment or cavity below the outer cuticular mantle (Figs. 1–7 and 12, *c*) is a site of ion accumulation in which considerable hydrostatic pressure develops before the salt solution is released through the cuticular pores. The possibility that considerable hydrostatic pressures develop within this compartment and that the salt solution is emitted through a valve-like cuticle and pressure release, lend credence to RUHLAND's (1915) observations that the emission was due to development of pressure within the glands. This suggestion coincides closely with the conclusion by ARISZ and associates (ARISZ et al., 1956; ARISZ, 1956) that secretion was dependent on an increase of turgor pressure within the gland. The only major difference with the concept of ARISZ and co-workers is that the pressure develops in the collecting compartment rather than in gland cells.

The idea of the possible involvement of symplastic transport and the underlying mesophyll cells in secretion, was originally developed by ARISZ (ARISZ et al., 1955; ARISZ, 1956) and further elaborated by LÜTTGE and his associates. Since this concept and the evidence for it have been recently reviewed in detail by LÜTTGE (1969, 1971) only a few major features will be covered in this chapter. As pointed out previously, according to this concept ion uptake occurs in the mesophyll cells. From there the ions move through the symplast to the glands where the salt is concentrated either in the vacuole, as in the bladder cells of *Atriplex* or, in the case of the *Limonium-Tamarix* types glands, the assumption is made that the ion concentration of the glands is regulated by exportation of salts from the cytoplasm of the secretory cells (LÜTTGE, 1971). In *Atriplex* supportive evidence comes from the observations that ion accumulation in the bladder cells is dependent on an energy source in the mesophyll (LÜTTGE and OSMOND, 1970; LÜTTGE et al., 1970) and that electrical transient changes in the bladders are directly related to changes in the mesophyll (LÜTTGE and OSMOND, 1970; LÜTTGE and PALLAGHY, 1969). The latter seems to indicate the absence of a membrane barrier between the bladder and mesophyll. Data from autoradiography studies suggest that in *Limonium*, ions are not concentrated in the glands to a level greater than that in the mesophyll (ZIEGLER and LÜTTGE, 1967). As LÜTTGE (1971) has pointed out, it is still not known where the active site(s) or pump(s) are located which are involved in the secretion phenomenon in the gland-leaf-mesophyll system. If, however, the major route of transport to the gland is via the apoplast, then the active site(s) must be associated with the glandular cells. The primary point of interest would be the plasmalemma of the inner gland cells along the transfusion zone, since the wall of the transfusion zone represents the end point of the apoplastic continuum. HILL has studied transit flux analyses of Na and Cl from an incubation medium to exudation in *Limonium* (HILL, 1970a). He reported that several compartments exist for each ion and that the transit half time of an ion was correlated with the filling of a compartment, which he concluded was cytoplasm, and that the gland directly controlled the ionic activity within this compartment. One of the major problems in these types of studies is clearly establishing the degree of coincidence or relationship between the filling of compartments in the gland and in the mesophyll such as the chloroplasts (LARKUM and HILL, 1970). If apoplastic transport does occur to the glands then much of HILL's data (HILL, 1970a, b; LARKUM and HILL, 1970) may indicate compartments in the glands rather than the mesophyll.

The ultrastructure of the salt glands of *Spartina* differs significantly from that described for dicotyledonous plants, and a different secretory mechanism has been suggested (LEVERING and THOMSON, 1971, 1972). Since the most unique feature of these glands is the numerous partitioning membranes enclosing extracellular channels in the basal cells, these investigators have suggested that these structures are involved in the secretory process. Wall protuberances occur in other salt glands (ZIEGLER and LÜTTGE, 1966; ATKINSON et al., 1967; THOMSON and LIU, 1967; SHIMONY and FAHN, 1968). Surface membrane elaborations which increase surface to volume ratios also occur in animal cells (cf. BERRIDGE and OSCHMAN, 1969), and as GUNNING and PATE (1969a, b) have pointed out, these types of cells are particularly efficient for absorption and secretion. During the past years DIAMOND and his associates (DIAMOND and TORMEY, 1966a, b; DIAMOND and BOSSERT, 1967, 1968) have developed the concept of a standing-gradient flow system to explain solute-linked water transport in epithelial cells of animals. DIAMOND and BOSSERT (1968) have pointed out that these cells are ... "constructed on a common geometrical plan: they possess long and narrow dead-end channels oriented parallel to the direction of fluid transport, open at the end towards which fluid is being transported, but closed at the end from which fluid is being absorbed". In the operation of the standing-gradient flow system solutes are pumped actively into the closed end of the extracellular channel forming a localized osmotic gradient which induces a passive flow of water. Subsequently the solute flows towards the open end of the channel and the fluid becomes progressively less hypertonic as more water enters along the channel. Dilations of the extracytoplasmic spaces during transport of ions and localization of salts in the spaces have been observed for animal tissues and provide experimental evidence for the standing osmotic gradient hypothesis (KAYE et al., 1966; TORMEY and DIAMOND, 1967; BERRIDGE and GUPTA, 1967; SCHMIDT-NIELSEN and DAVIS, 1968).

The following hypothesis based on the ultrastructure and organization of the salt glands of *Spartina* and a comparison of these features with apparently analogous structures and systems in animal tissues, was proposed for the mechanism of salt secretion in *Spartina* salt glands (LEVERING and THOMSON, 1971, 1972). They suggested that there is an active secretion of ions into the extracellular channels of the partitioning membranes in the basal cells. This is probably accomplished with the utilization of energy supplied by the associated mitochondria. (Interestingly enough a similar association of mitochondria and membranes has been observed in the salt secreting glands of the brine shrimp and COPELAND (1966) has postulated that these arrangements are integral units involved in the transport of ions across the plasma membrane). If a coupled solute-water transport system is assumed (DIAMOND and BOSSERT, 1967), water will move passively through the basal cell and into the extracellular channels along an osmotic gradient.

LEVERING and THOMSON (1971) observed dilations of the extracellular channels of the partitioning membranes in some glands. In subsequent experiments they found that the channels were highly dilated in glands of leaves placed in NaCl but the partitioning membranes were closely appressed in glands of leaves treated with distilled water (LEVERING and THOMSON, 1972). Further studies were done using the potassium pyroantimonate technique of KOMNICK (1962) for the localization of Na at the ultrastructural level. In these studies heavy deposits of electron dense reaction product were observed in the channels and dilations of the partitioning membranes (LEVERING and THOMSON, 1972). The results of these experiments are quite comparable to those with absorptive and secretory cells in animals and provide evidence that the extracellular channels are sites of ion accumulation, which supports the standing-gradient hypothesis.

Thus it appears that the mechanism of secretion of *Spartina* is almost identical to fluid transporting systems in animals. The possibility that the standing-gradient flow system is operative in other absorptive and secretory tissues and cells in plants is likely and should be further explored. Structural similarities also exist between salt glands and other types of plant glands and secretory cells. The most distinctive similarity is the occurrence of numerous wall protuberances which effectively increase the surface to volume ratio of the secretory cells (LÜTTGE, 1966; SCHNEPF, 1964, 1965, 1969; THOMSON and LIU, 1967; ATKINSON et al., 1967). Similar wall protuberances occur in a variety of other tissues such as developing embryos (JENSEN, 1963) between adjacent cells of sporophyte and gametophyte tissue (KELLEY, 1969), and in transfer cells (GUNNING and PATE, 1969a, b). GUNNING and PATE reasoned that such cells are likely to be efficient in either absorptive or secretory functions and suggested that the ingrowths provide relatively static compartments from which solutes could be absorbed or secreted. This concept is somewhat similar to the standing-gradient hypothesis proposed by DIAMOND. In regard to salt glands, additional support for this hypothesis comes from the fact that according to the standing-gradient hypothesis, the rate of secretion is independent of the concentration of the fluid and the secreted fluid may be isotonic with the underlying tissue. This would explain ARISZ et al. (1955) observation that the concentration of the secreted salt is independent of rate of secretion and the observations of ATKINSON et al. (1967) that, in mangroves, the water potential of the secreted fluid is close to that of the leaf. Although the salt glands of some plants do not have long membranous channels, such as in *Spartina* and some lack or have few wall protuberances, the large collecting compartments between the cuticle and gland cells very possibly function as a site of ion accumulation and osmotic uptake of water.

Some enzymes, particularly ATPases, are known to play an important role in ion transport in animal systems (SKOU, 1964), particularly in absorptive and secretory cells (BERRIDGE and GUPTA, 1968; ERNST and PHILPOTT, 1970). There is growing evidence of their importance in ion transport in plants (FISHER and HODGES, 1969; KYLIN and GEE, 1970), and recently JENNINGS (1968) has developed the concept that these ion stimulated (particularly sodium) ATPases, i.e. sodium pumps, have a key function in halophytes and in salt resistance. However, much more information on the roles of these enzymes in salt secretion and ion transport in plants is needed before their importance can be assessed completely. If the assumption is made that they are important, then their localization in the gland relative to the wall protuberances, partitioning membranes and the plasmalemma along the wall of the transfusion zone would add considerable insight into the mechanism of secretion.

Acknowledgements and Comments

Since many aspects of secretion and salt glands have been reviewed recently in detail by SCHNEPF (1969) and LÜTTGE (1969, 1971), the topic and review of the literature was almost entirely constricted to various aspects of salt glands. The reader, if interested in the broader aspects of secretion, should consult the above reviews. In preparing this review the author must express his appreciation for the excellent technical assistance of KATHRYN PLATT. Acknowledgement of support from the National Science Foundation grant (GB 8199) for much of the original studies presented herein is also made.

References

ARISZ, W. H.: Significance of the symplasm theory for transport across the root. Protoplasma **46**, 5–62 (1956).

ARISZ, W. H., CAMPHUIS, I. J., HEIKENS, H., VAN TOOREN, A. J.: The secretion of the salt glands of *Limonium latifolium* ktze. Acta Botan. Neerl. **4**, 322–338 (1955).

ATKINSON, M. R., FINDLAY, C. P., HOPE, A. B., PITMAN, M. G., SADDLER, H. D. W., WEST, K. R.: Salt regulation in the mangroves *Rhizophora mucronata Lam.* and *Aegialitis annulata R.* Br. Australian J. Biol. Sci. **20**, 589–599 (1967).

BERRIDGE, M. J., GUPTA, B. L.: Fine-structural changes in relation to ion and water transport in the rectal papillae of the blowfly *Calliphora*. J. Cell Sci. **2**, 89–112 (1967).

BERRIDGE, M. J., GUPTA, B. L.: Fine-structural localization of adenosine triphosphatase in the rectum of *Calliphora*. J. Cell Sci. **3**, 17–32 (1968).

BERRIDGE, M. J., OSCHMAN, J. L.: A structural basis for fluid secretion by Malpighian tubules. Tissue and Cell. **1**, 247–272 (1969).

BERRY, W. L.: Characteristics of salts secreted by *Tamarix aphylla*, Am. J. Botany **57**, 1226–1230 (1970).

BERRY, W. L., THOMSON, W. W.: Composition of salt secreted by salt glands of *Tamarix aphylla*. Can. J. Botany **45**, 1774–1775 (1967).

BLACK, R. F.: Leaf anatomy of Australian members of the genus *Atriplex*. I. *Atriplex vesicaria* Heward and *A. nummularia* Lindl. Australian J. Botan. **2**, 259–286 (1954).

BONNETT, H. T., JR.: The root endodermis: fine structure and function. J. Cell Biol. **37**, 109–205 (1968).

CAMPBELL, C. J., STRONG, J. E.: Salt gland anatomy in *Tamarix pentandra (Tamaricaceae)*. Southwest Nat. **9**, 232–238 (1964).

COPELAND, E.: Salt transport organelle in *Artemia salenis*. Science **151**, 470–471 (1966).

DIAMOND, J. M., BOSSERT, W. H.: Standing-gradient osmotic flow. A mechanism for coupling of water and solute transport in epithelia. J. Gen. Physiol. **50**, 2061–2083 (1967).

DIAMOND, J. M., BOSSERT, W. H.: Functional consequences of ultrastructural geometry in "backwards" fluid-transporting epithelia. J. Cell Biol. **37**, 694–702 (1968).

DIAMOND, J. M., TORMEY, J. MCD.: Role of long extracellular channels in fluid transport across epithelia. Nature **210**, 817–820 (1966 a).

DIAMOND, J. M., TORMEY, J. MCD.: Studies on the structural basis of water transport across epithelial membranes. Federation Proc. **25**, 1458–1463 (1966 b).

ERNST, S. A., PHILPOTT, C. W.: Preservation of Na-K activated and Mg-activated adenosine triphosphatase activity of avian salt gland and telost gill with formaldehyde as fixative. J. Histochem. Cytochem. **18**, 251–263 (1970).

ESAU, K.: Plant Anatomy, 2nd ed. pp. 767. New York-London-Sidney: John Wiley and Sons 1965.

FINDLAY, N., MERCER, F. V.: Nectar Production in *Abutilon*. I. Movement of Nectar through the Cuticle. J. Biol. Sci. **24**, 647–56 (1971 a).

FINDLAY, N., MERCER, F. V.: Nectar Production in *Abutilon*. II. Submicroscopic Structure of the Nectary. J. Biol. Sci. **24**, 657–64 (1971 b).

FISHER, J., HODGES, T. K.: Monovalent ion stimulated adenosine triphosphatase from oat roots. Plant Physiol. **44**, 385–395 (1969).

GUNNING, B. E. S., PATE, J. S.: "Transfer cells"—Plant cells with wall ingrowths, specialized in relation to short transport of solutes—their occurrence, structure, and development. Protoplasma **68**, 107–133 (1969 a).

GUNNING, B. E. S., PATE, J. S.: Vascular transfer cells in angiosperm leaves a taxonomic and morphological survey. Protoplasma **68**, 135–156 (1969 b).

HABERLANDT, G.: Physiological plant anatomy, 777 pp. London: MacMillan and Co., Ltd. 1914.

HELDER, R. J.: The loss of substances by cells and tissues (salt glands). In: RUHLAND, W. (Ed.): Handbuch der Pflanzenphysiologie, Vol. 2, pp. 468–488. Berlin-Göttingen-Heidelberg: Springer 1956.

HILL, A. E.: Ion and water transport in *Limonium*. I. Active transport by the leaf gland cells. Biochim. Biophys. Acta **135**, 454–460 (1967 a).

HILL, A. E.: Ion and water transport in *Limonium*. II. Short-circuit analysis. Biochim. Biophys. Acta **135**, 461–465 (1967 b).

HILL, A. E.: Ion and water transport in *Limonium*. III. Time constants of the transport system. Biochim. Biophys. Acta **196**, 66–72 (1970 a).

HILL, A. E.: Ion and water transport in *Limonium*. IV. Delay effects in the transport process. Biochim. Biophys. Acta **196**, 73–79 (1970 b).

JENNINGS, D. H.: Halophytes, succulence and sodium in plants—a unified theory. New Phytol. **67**, 899–911 (1968).

JENSEN, W. A.: Cell development during plant embryogenesis. Brookhaven Symp. Biol. **16**, 179–202 (1963).

KAYE, G. I., WHEELER, H. O., WHITLOCK, R. T., LANE, N.: Fluid transport in the rabbit gallbladder. Jour. Cell. Biol. **30**, 237–268 (1966).

KELLEY, C.: Wall projections in the sporophyte and gametophyte of *Sphaerocarpus*. J. Cell Biol. **41**, 910–914 (1969).

KOMNICK, H.: Electronenmikroskopische Lokalisation von Na^+ und Cl^- in Zellen und Geweben. Protoplasma **55**, 414–418 (1967).

KYLIN, A., GEE, R.: Adenosine triphosphatase activities in leaves of the mangrove *Avicennia nitida Jacq*. Plant Physiol. **45**, 169–172 (1970).

LARKUM, A. W. D., HILL, A. E.: Ion and water transport in *Limonium*. V. The ionic status of chloroplasts in the leaf of *Limonium vulgare* in relation to the activity of the salt glands. Biochim. Biophys. Acta **203**, 133–138 (1970).

LEVERING, C. A., THOMSON, W. W.: The ultrastructure of the salt gland of *Spartina foliosa*. Planta **97**, 183–196 (1971).

LEVERING, C. A., THOMSON, W. W.: Studies on the ultrastructure and mechanism of secretion of the salt gland of the grass *Spartina*. Proc. 30th Electron Microscope Soc. of America, 222–223 (1972).

LÜTTGE, U.: Funktion und Struktur pflanzlicher Drüsen. Die Naturwissenschaften **53**, 96–103 (1966).

LÜTTGE, U.: Aktiver Transport (Kurzstreckentransport bei Pflanzen). Protoplasmatologia **8**, 1–146 (1969).

LÜTTGE, U.: Structure and function of plant glands. Ann. Rev. Plant Physiol. **22**, 23–44 (1971).

LÜTTGE, U., OSMOND, C. B.: Ion absorption in *Atriplex* leaf tissue. III. Site of metabolic control of light-dependent chloride secretion to epidermal bladders. Australian J. Biol. Sci. **23**, 17–25 (1970).

LÜTTGE, U., PALLAGHY, C. K.: Light triggered transient changes of membrane potentials in green cells in relation to photosynthetic electron transport. Z. Pflanzenphysiol. **61**, 58–67 (1969).

LÜTTGE, U., PALLAGHY, C. K., OSMOND, C. B.: Coupling of ion transport in green cells of *Atriplex spongiosa* leaves to energy sources in the light and in the dark. J. Membrane Biol. **2**, 17–30 (1970).

MACROBBIE, E. A. C.: Fluxes and compartmentation in plant cells. Ann. Rev. Plant Physiol. **22**, 75–96 (1971).

METCALFE, C. R., CHALK, L.: Anatomy of the dicotyledons, Vol. I and II, 1500 p. Oxford: Clarendon Press 1950.

MOZAFAR, A., GOODIN, J. R.: Vesiculated hairs: A mechanism for salt tolerance in *Atriplex halimus* L. Plant Physiol. **45**, 62–65 (1970).

OSMOND, C. B., LÜTTGE, U., WEST, K: R., PALLAGHY, C. K., SHACHER-HILL, B.: Ion absorption in *Atriplex* leaf tissue. II. Secretion of ions to epidermal bladders. Australian J. Biol. Sci. **22**, 797–814 (1969).

POLLACK, G., WAISEL, Y.: Salt secretion in *Aeluropus litoralis* (Willd.) Parl. Ann. Botan. **34**, 879–888 (1970).

RUHLAND, W.: Untersuchungen über die Hautdrüsen der Plumbaginaceen. Ein Beitrag zur Biologie der Halophyten. J. Wiss. Botan. **55**, 409–498 (1915).

SCHMIDT-NIELSEN, K.: Physiology of salt glands. In: WOHLFAHRT-BOTTERMAN, K. E. (Ed.): Sekretion and exkretion, pp. 269–288. Berlin-Heidelberg-New York: Springer 1965.

SCHMIDT-NIELSEN, B., DAVIS, L. E.: Fluid transport and tubular intercellular spaces in reptilian kidneys. Science **159**, 1105–1108 (1968).

SCHNEPF, E.: Über Zellwandstrukturen bei Köpfchendrüsen der Schuppenblätter von *Lathraea clandestina* L. Planta **60**, 473–482 (1964).

SCHNEPF, E.: Licht- und elektronenmikroskopische Beobachtungen an den Trichom-Hydathoden von *Cicer arietinum*. Z. Pflanzenphysiol. **53**, 245–254 (1965).

SCHNEPF, E.: Sekretion und Exkretion bei Pflanzen. Protoplasmatologia. **8**, 1–181 (1969).

SCHOLANDER, P. F.: How mangroves desalinate seawater. Physiol. Plantarum **21**, 251–261 (1968).

SCHOLANDER, P. F., BRADSTREET, E. D., HAMMEL, H. T., HEMMINGSEN, E. A.: Sap concentration in halophytes and some other plants. Plant Physiol. **41**, 529–532 (1966).

SCHOLANDER, P. F., HAMMEL, H. T., BRADSTREET, E. D., HEMMINGSEN, E. A.: Sap pressure in vascular plants. Sci. **148**, 339–345 (1965).

SCHOLANDER, P. F., HAMMEL, H. T., HEMMINGSEN, E., GARRY, W.: Salt balance in mangroves. Plant Physiol. **37** (6), 722–729 (1962).

SCHTSCHERBACK, J.: Über die Salzausscheidung durch die Blätter von *Statice gmelini*. Ber. Deut. Botan. Ges. **28**, 30–34 (1910).

SHACHAR-HILL, B., HILL, A. E.: Ion and water transport in *Limonium*. Biochim. Biophys. Acta **211**, 313–317 (1970).

SHIMONY, C., FAHN, A.: Light and electron microscopical studies on the structure of salt glands of *Tamarix aphylla* L. J. Linn. Soc. **60**, 283–288 (1968).

SKELDING, A. D., WINTERBOTHAM, J.: The structure and development of the hydathodes of *Spartina townsendii* groves. New Phytol. **38**, 69–79 (1939).

SKOU, J. C.: Enzymatic aspects of active linked transport of Na^+ and K^+ through the cell membrane. Progr. Biophys. Molec. Biol. **14**, 131–166 (1964).

SMAOUI, M. A.: Differentiation des trichomes chez *Atriplex halimus L.* C. R. Acad. Sci. **273**, 1268–1271 (1971).

STOCKING, C.: Guttation and bleeding. In: RUHLAND, W. (Ed.): Handbuch der Pflanzenphysiologie, Vol. 3, pp. 489–502. Berlin-Göttingen-Heidelberg: Springer 1956.

THOMSON, W. W., LIU, L. L.: Ultrastructural features of the salt gland of *Tamarix aphylla* L. Planta **73**, 201–220 (1967).

THOMSON, W. W., BERRY, W. L., LIU, L. L.: Localization and secretion of salt by the salt glands of *Tamarix aphylla*. Proc. Natl. Acad. Sci. U.S. **63**, 310–317 (1969).

TORMEY, J. MCD., DIAMOND, J. M.: The ultrastructural route of fluid transport in rabbit gall bladder. J. Gen. Physiol. **50**, 2031–2060 (1967).

VOLKEN, G.: 1884. Die Kalkdrüsen der Plumbagineen. Ber. Deut. Botan. Ges. **2**, 334–342 (1967).

WAISEL, Y.: Ecological studies on *Tamarix aphylla* (L.) Karst. III. The salt economy. Plant Soil. **13**, 356–364 (1961).

ZIEGLER, H., LÜTTGE, U.: Die Salzdrüsen von *Limonium vulgare*. I. Die Feinstruktur. Planta **70**, 193–206 (1966).

ZIEGLER, H., LÜTTGE, U.: Die Salzdrüsen von *Limonium vulgare*. II. Die Lokalisierung des Chloride. Planta **74**, 1–17 (1967).

Chapter VIII

Metabolic and Biochemical Aspects of Salt Tolerance

A. Kylin and R. S. Quatrano

A. General Introduction

It seems a truism to say that every organism must have its biochemical and structural properties adapted so that it is able to function in its habitat. Nevertheless, when investigating the causal relationships between ecological conditions and biochemical responses, one ends up with information relevant to our understanding of the regulation of basic cell functions, such as membranes and metabolism. Apart from the immediate importance of studying major ecological problems such as salinity from the biochemical point of view, it should be possible to utilize the data to shed light on fundamental theoretical aspects of life.

One may expect adaptability in plants to be founded on biochemical properties to a higher degree than in animals, both as regards the genetically fixed potential to survive in a specific habitat and as regards the phenotypical variability, which determines the range of habitats and conditions that are open for a species. Unlike most animals, plants are normally stationary and cannot move away when they meet adverse conditions. At the same time, plants must cope with the external medium in a much more direct manner than animals, where more or less well regulated body fluids can serve as the immediate medium for the cells and tissues.

Whatever the value of the above general outlook, it has so far led only to limited and scattered work. The approach of plant ecologists has traditionally been sociological or autecological—that is, from the standpoint of physiology their concern has been mainly with whole plants or whole organs. The interest of biochemistry has mainly been turned towards man or animals, comparatively little towards plants. The purpose of the present review must be to encourage research, not to make an attempt at final conclusions. The scattered data that are available will be connected with the help of general biochemical hypotheses, which are not necessarily substantiated within the field of salt resistance of plants, in the hope that this will indicate possible approaches and encourage new experiments.

Salinity is a complex phenomenon in itself. One major aspect is the lowering of the osmotic potential. This is equivalent to a lowering of the water activity, a property which is held in common with stress situations exerted by frost or by dryness. Superimposed on this are the specific effects of the various ions concerned. Chlorides, sulphates or carbonates may predominate among the anions of a soil, depending on the circumstances (Strogonov, 1964). Among the salt tolerant plants, one may find chloride-halophytes, sulphate-halophytes and alkali-halophytes, where the cations are mainly compensated for by oxalate or organic acids, not by inorganic anions (WALTER and KREEB, 1970). The dominant cation of a saline soil is mostly sodium, but calcium or combinations of calcium and magnesium also occur, especially at the later stages of soil development (DAUBENMIRE, 1959). These are background facts that should be borne in mind in the following.

B. Effects of Salinity on the Activity of Isolated Enzymes, and on the Formation of Enzymes

The hydration and the conformational state of proteins and of other cytoplasmatic constituents are affected by the osmotic potential and by the specific composition of the medium. This corresponds to the well-known fact that the osmotic potential and ionic composition of the medium are important in work with subcellular preparations and enzymes (LATIES, 1954; HACKETT, 1961; to mention some examples concerning higher plants). The theory of some of the possible interactions may be found in works of KLOTZ (1958), EISENMAN (1960), KAVANAU (1964), STEINHARDT and BEYCHOK (1964), EVANS and SORGER (1966), VON HIPPEL and SCHEICH (1969); SUELTER (1970), and KUIPER (1972b).

It is an old experience that the osmotic factor affects various cell functions, so that at water saturation and at high water deficits their rate is lower than at some optimal water conditions (MOTHES, 1956; STOCKER, 1956). At the same time, the optimal water conditions may vary, not only with the organism but also with the kind of tissue within a given organism (M. and R. BOUILLENNE-WALRAND, 1926), as well as with the osmotic agent used (TAKAOKI, 1957).

Against this background it is not surprising that bacterial enzymes isolated from species with different halotolerance show different optima when tested against salt concentration *in vitro* (BAXTER and GIBBONS, 1954). Contrariwise, when enzymes from higher plants have been tested for salt resistance *in vitro*, the enzymes from halophytes have not differed to any great extent from those of glycophytes (GREENWAY and OSMOND, 1972; FLOWERS, 1972a, b). It was even reported that phosphoenolpyruvate carboxylase from salt resistant *Atriplex spongiosa* was more salt sensitive *in vitro* than the same enzyme from *Zea mays*, one of the most salt sensitive species existing (OSMOND and GREENWAY, 1972). Similarly, enzymes prepared from members of the highly salt tolerant algal genus *Dunaliella* are inhibited *in vitro* by salt concentrations far below those of the growth media (JOHNSON et al., 1968; JOKELA, 1969; BEN-AMOTZ and AVRON, 1972). It is unclear whether the similarity *in vitro* between enzymes from halophytes and those from glycophytes is due to preparation artifacts or to different abilities of halophytes and glycophytes to regulate the ion concentration around their enzymes *in vivo* (GREENWAY and OSMOND, 1972). Anyway, the rule is not without exceptions: in the investigation of FLOWERS (1972b) there were marked differences between species in the salt sensitivities of the malic dehydrogenases and the ATPases.

The dominant results become different from the above, when one investigates the adaptative differences that occur when a plant species is cultivated at different levels of salinity. It is true that WEIMBERG (1970) did not find any significant differences in the levels of 18 different enzymes from pea seedlings (cultivar Alaska) grown either in a liquid medium or in the same medium salinized with 5 atmospheres of NaCl, KCl, Na_2SO_4 or K_2SO_4. However, using root tips of another cultivar of pea (Laxton Progress), the group of POLJAKOFF-MAYBER investigated the effect of a range of salinities, either as NaCl or as Na_2SO_4 (PORATH and POLJAKOFF-MAYBER, 1964, 1968). The changes that occurred led to an increase in the pentose-phosphate pathway under salinity caused by NaCl, whereas Na_2SO_4 had little effect in this respect.

Continued investigations with pea of the cultivar Laxton Progress demonstrated that when the plants are grown in the presence of NaCl, a new isozyme of malate dehydrogenase is developed in addition to the two that are found in non-stressed plants and in plants given Na_2SO_4 (HASSON-PORATH and POLJAKOFF-MAYBER, 1969). Lactic dehydrogenase coupled to NADH is depressed by applications of NaCl to the growth medium and increased by treatments with Na_2SO_4, whereas the activity linked to NADPH is acted upon in the reverse (HASSON-PORATH and POLJAKOFF-MAYBER, 1970). Differential effects were also found on phosphatases and on oxidative phosphorylation (HASSON-PORATH and POLJAKOFF-MAYBER, 1971).—Ribulosediphosphate carboxylase and phosphoenolpyruvate carboxylase were affected by applications of NaCl to *Zea mays*, *Chloris gayana* and *Aeluropus litoralis* in a way that depended on the species, on the age of the leaf and on the amount of NaCl (SHOMER-ILAN and WAISEL, 1973). Likewise, the activity of extracted acid phosphatase decreased while that of phosphoenolpyruvate phosphatase increased, when *Suaeda fruticosa* was given extra NaCl in the medium (AHMAD and HEWITT, 1971).

The ATPases from a mainly microsomal fraction of root homogenates of wheat and oat vary in a way which is dependent on the salt level of the growth medium and which can be correlated with experience from cultivation in the field (KYLIN and KÄHR, 1973 and unpublished). Oat gives a good crop on acid soils, which are low in calcium, whereas wheat needs fair amounts of calcium. The ATPase from oat is dominated by activation through magnesium, and only a minor part is activated by calcium. In wheat, activation by calcium is more obvious than activation by magnesium—a finding which could kinetically be resolved as an interaction between two sites, one where calcium and magnesium both activate although they compete, and another where magnesium blocks the system but is outcompeted by calcium so that the block is relieved.

Furthermore, acidity is often a sign that a soil has a generally low salt status, whereas high calcium is correlated with a good salt status of the soil. Oat, which grows well on acidic, low salt soils, gave much more ATPase activity when the plants were cultivated on a low salt medium containing only 0.2 mM $CaSO_4$, than when the plants were given a complete nutrient solution. Wheat, which prefers calcium-containing soils of a good salt status, behaves in the opposite direction. In wheat, the change in ATPase activity is due to a qualitative change in the proteins in relation to the medium given, since both treatments gave roots with the same amount of protein per g fresh weight.

Summarizing, one cannot *a priori* expect an accidentally chosen enzyme to be salt resistant *in vitro* or to change its activity *in vivo* in response to salt stress just because one is investigating a halophyte. At the same time responses do occur and correlations are found. A reasonable approach may then be to select more complex cell functions of such a character that they can be assumed to have a function in connection with salt resistance. The following subchapters will be devoted to some areas of this type.

C. Cellular Functions and Properties Related to Salt Tolerance

I. Osmoregulation

Osmoregulation for maintenance of turgor under salt stress can *a priori* be regarded as a necessity for halotolerant cells. The extremes in this case may be represented on the one side by the species of Halobacterium, which GINZBURG (1969) found to be completely permeable for molecules up to the size of polyvinylpyrrolidone (M.W. 30–40000)—thus solving the problem by avoiding it. On the other side, higher plants are known to utilize ion transport (WAISEL, 1972), which is to a large extent active and coupled to metabolism.

Intermediate cases occur. NORKRANS and KYLIN (1969) made a study of the balance of Na and K in the marine yeast *Debaryomyces hansenii* as compared to non-halotolerant baker's yeast, *Saccharomyces cerevisiae*. At 4% (0.68 M) or more NaCl in the medium, neither organism had a corresponding internal content of salt. Nevertheless, *Debaryomyces* lost only a little water when transferred to media of high salinity (up to 16% NaCl was tested) and regained volume quickly. In a similar situation, *Saccharomyces* lost much more water and regained it only slowly. One of the capacities that make *Debaryomyces* halotolerant appears to be the ability of rapid osmoregulation with the help of rather small organic molecules produced by metabolism.

Polyalcohols have been suggested as part of an osmoregulation independent of ion transport (ONISHI, 1963). This fits well with the situation in the highly salt resistant algal genus *Dunaliella*. JOKELA (1969) found that the ion content could not account for the maintenance of cell volume in *Dunaliella tertiolecta*. Instead, rapid variations in the contents of glycerol occurred. In *Dunaliella parva* the changes in glycerol have been quantitatively shown to be of sufficient size for osmoregulation (BEN-AMOTZ and AVRON, 1973). Its formation is assumed to take place via photosynthesis or via metabolic degradation of starch, depending on the light conditions. A similar mechanism at a much lower salt level has been demonstrated in the fresh-water *Ochromonas* (KAUSS, 1967), and may exist also in *Chlamydomonas* (OKAMOTO and SUZUKI, 1964).

As mentioned earlier, enzymes from bacteria correlate with the salt tolerance of the species from which they are isolated (BAXTER and GIBBONS, 1954), whereas enzmyes isolated from *Dunaliella* are salt sensitive despite the halotolerance of the organism (JOHNSON et al., 1968; JOKELA, 1969; BEN-AMOTZ and AVRON, 1972). Preliminary results indicate that enzymes from *Dunaliella* cells are resistant to glycerol or require glycerol for full enzymic activity (BEN-AMOTZ and AVRON, 1973). This effect of the main osmoregulatory substance in *Dunaliella* is interesting, since it is often assumed that the effect of glycerol as a protectant against plasmatic damage during freezing is due to the formation of a non-freezing mixture polyalcohol-water-salt, which prevents the salt damage on proteins which would otherwise occur when water "freezes out" (SMITH, 1954; MAZUR, 1969).—By the same token, one of the main responses of *Chlorella* subjected to low water potentials is a rapid increase in the synthesis of sucrose (HILLER and GREENWAY, 1968). Older literature on higher plants is conflicting and the experiments were made on a long-time basis, so that the information is difficult to interpret: too many functions are integrated over too long periods. For access to this literature, the reader is referred to HILLER and GREENWAY (1968).

II. Physiology of Membranes

1. Membrane Function

a) General Considerations. The cell membranes are important in the response of the cells to various types of solutes. The effects of an osmotic treatment depend both on the type of tissue and on the osmoticum used. In experiments by GREENWAY (1970) it was demonstrated that non-vacuolated cells from root tips of *Zea mays* were less severely affected by a sequence of plasmolysis-deplasmolysis than cells from the vacuolated parts of the root. Low water potentials were more deleterious when they were caused by non-permeating mannitol than when obtained by quickly penetrating ethylene glycol (GREENWAY and LEAHY, 1970).

Different substances can be released from cells dependent upon the species and upon the treatment, and such release can change the cell functions. NIEMAN (1969) treated salt sensitive carrot with NaCl, and found that the tissue released polysaccharides, lipids, and three different proteins. A graded treatment could be performed, so that only two of the proteins were released: this was accompanied by suppression of the uptake of glucose and phosphate without immediate effects on cell permeability. Tissues from salt tolerant beet resisted the treatment. Enzymatic proteins with α-glucosidase activity can be released from the surface of the cells of callus from *Convolvolus* by salt treatment (KLIS, 1971).

Ion transport through the cell membranes is partly an active process, requiring energy. Complete energy conversion is found only in mitochondria and chloroplasts. The question arises how the plasmalemma, the tonoplast and other membranes lacking energy production of their own are energized for active transport of ions. The obvious answer is ATP or equivalent sources of energy. Translated to biochemistry in the test-tube, one would expect the system for active transport of ions to be detectable as an ATPase specifically activated by the ion transported. So far, the concept is nothing that has to be "proved"—it is inherent in our knowledge of the cell and the way it functions. What has to be proved is whether the properties of a given ATPase are close enough to the properties of a given type of transport, so that one may assume that the ATPase is really an expression of the transport system. This was first done for the coupled transport of sodium and potassium studied as ($Na^+ + K^+$)-stimulated ATPases, as reviewed by SKOU (1964). Once this is done, the field is open to ask questions about the biochemical composition that enables the system to work, and about the modifications imposed by ecological specialization and adaptation.

One of the practical problems that have to be solved by the biochemistry of a system for ion transport, is how to get such a hydrophilic unit as an ion through the lipophilic parts of a membrane. The existence of specific carrier proteins was shown by PARDEE (1968). To the extent that the transport is active, this "carrier site" can be coupled to another protein with the function of energy conversion (PARDEE, 1968)—an "ATPase site". The conformation and function of the proteins may be influenced by the associated lipids (BENSON, 1964; GREEN and TZAGOLOFF, 1966). Not only the active but also the passive transport can be controlled by the lipids of the membrane. This has been shown in model systems for the permeability of both water (GRAZIANI and LIVNE, 1972) and salts (HOPFER et al., 1970; PAPAHADJOPOULOS, 1971).

b) ($Na^+ + K^+$)-Activated ATPases. The interest of the present author and his collaborators arose from the demonstration of a mechanism for the extrusion of sodium from the microalga *Scenedesmus* (KYLIN, 1964), which showed properties that could be expected if a biochemical system like the ($Na^+ + K^+$)-activated ATPases was responsible (KYLIN, 1966). At the time the mechanism was controversial, since most of the groups concerned with ion transport in plants maintained that the rate-limiting and regulating steps were to be found only in the uptake processes, not in extrusion. Electrophysiological evidence for extrusion was available, but could also be interpreted as reflecting only different permeabilities for different ionic species (DAINTY, 1962). Our feeling was that biochemical evidence was needed from plants, of the same type that had been produced for the animal systems (SKOU, 1964).

Scenedesmus is not an ideal material for subcellular preparations, due to its hard cell walls. We started, instead, from an ecological analysis of the situation. The physiological role of extrusion of the highly hydrated sodium ion is to keep the cytoplasm free from damage by overhydration. The mechanism is presumably more important for naked cells than for cells with a cell wall, where wall pressure may fill the same function as extrusion of sodium (ROTHSTEIN, 1964). Evidently, the extrusion mechanism must have a special selection value for halophytes.

This way of reasoning led us to sugar beet, which has been developed from the halophyte *Beta maritima*. As an agricultural plant, it has the additional advantage that its relation to mineral nutrition is well known. A remarkable feature in this respect, which encouraged us further in our choice, is that sodium in the fertilizer will lead to an increased crop of the sugar beet, and that lack of sodium will lead to symptoms similar to those of potassium deficiency (EL-SHEIK et al., 1967).

The ecological approach proved successful, and a ($Na^+ + K^+$)-activated ATPase system could be demonstrated in a mainly microsomal fraction of roots of sugar beet seedlings (HANSSON and KYLIN, 1969). Following the same ecological line of thought, leaves of the mangrove *Avicennia nitida* have salt-excreting glands, and contain ($Na^+ + K^+$)-stimulated ATPase activity (KYLIN and GEE, 1970).

HALL (1971) demonstrated by electrophoresis that several ion activated ATPases were present in barley roots. The ($Na^+ + K^+$)-activated ATPase from sugar beet has not yet been similarly characterized. However, it is homogenous enough to perform analyses that give good fit to kinetic models that have been analyzed in animal systems, for instance by HEXUM et al. (1970). The primary substrate appears to be the MgATP-complex, with free ATP acting as a competitive inhibitor but little influenced by free Mg. Sodium and potassium cooperate with MgATP in a mainly non-competitive manner (S. LINDBERG, unpublished).

At this stage a point of definition should be noted. In order to be critically called a ($Na^+ + K^+$)-stimulated ATPase, the system must be shown to give optimal activity at specific ratios of sodium to potassium. It is thus not enough to test only sodium and only potassium and find that both ions are stimulating—this can be due to an unspecific stimulation of monovalent cations. Furthermore, the ratios must be tested at a constant concentration of combined ($Na^+ + K^+$), otherwise a stimulation or inhibition measured may be due to a change in ionic strength. These precautions are related to general protein chemistry (cf. Section B of this chapter). In practice they mean that the system must be tested with densely spaced ratios of Na:K at several different levels of ionic strength (HANSSON and KYLIN, 1969; KYLIN and GEE, 1970).

With regard to the above definition, most of the older work in the field is insufficiently substantiated and of less interest for the present purposes. The reader is referred to the paper by KYLIN (1973) for a review. The investigations of LAI and THOMPSON (1971, 1972a; *Phaseolus vulgaris*) BOWLING et al. (1972; conducting bundles of *Beta vulgaris* and *Heracleum sosnovskyi*) and of KRASAVINA and VYSKREBENTSEVA (1972, gourd) are interesting, since they contain other types of information indicating that they may be concerned with ($Na^+ + K^+$)-activated ATPases related to transport.

c) Correlations between Properties of Membranes and Isolated ATPases. As pointed out above, one of the crucial questions in work with ion stimulated ATPases as transport systems is to obtain enough correlations to make it likely that one is working with a transport system and not with something which is "just an ATPase". The ecological approach as discussed is indirect evidence in itself but more is needed. Further correlations were obtained by the use of a series of inbred strains of sugar beet with different properties as salt accumulators in the field (KYLIN and HANSSON, 1971).

The one extreme is a (low salt, high potassium) strain and shows one major ($Na^+ + K^+$)-stimulated ATPase activity, where high concentrations of potassium are needed for optimal activity, and modification due to sodium is rather small. The other extreme is a (high salt, high sodium) strain, also with one major ($Na^+ + K^+$)-stimulated ATPase, but this time working optimally at low concentrations of potassium and highly modified by sodium in the assay medium. The other two strains are intermediate in their salt relations, and each of them shows two different optima for ($Na^+ + K^+$)-stimulated ATPase, in the same way as found for the normal field beets by HANSSON and KYLIN (1969).

Apart from giving further correlations between the ATPase and salt accumulation/ion transport, the data have a bearing on halophytism in itself. Salt tolerant, non-vacuolated cells have a high ratio of K:Na when they are compared with non-tolerant cells, as shown by NORKRANS and KYLIN (1969) for two types of yeast in situations of salt stress. This is expected, since one of the functions behind salt tolerance is a well-developed mechanism in the plasmalemma for extrusion of Na in exchange for K (ROTHSTEIN, 1964). Nevertheless, higher plants with vacuolated cells have a low ratio of K:Na in halotolerant species as compared to glycophytes (COLLANDER, 1941).

The apparent paradox can be resolved if the single $(Na^+ + K^+)$-stimulated ATPase of the (low salt, high potassium) strain of sugar beet is localized in the plasmalemma; and the one of the (high salt, high sodium) strain in the tonoplast. The more viable intermediate strains, as well as the normal field beets would have one transport system taking sodium from the cytoplasm out to the medium, and another transport system taking sodium from the cytoplasm in to the vacuole. In the vacuole, sodium is not only harmless but of actual value to maintain the turgor pressure of the cells under salt stress, at the same time as the double transport system would ensure optimal relief for the cytoplasm. With the major part of the sodium in the vacuole, the data of COLLANDER (1941) no longer contradict the data for non-vacuolated cells.

The hypothesis is confirmed by data of electrochemistry and flux measurements (SPANSWICK and WILLIAMS, 1964; SPANSWICK et al., 1967; PIERCE and HIGINBOTHAM, 1970). It is true that WAISEL and ESHEL (1971) found more sodium in the "cytoplasm" than in the vacuole of *Suaeda monoica* when investigated by X-ray micro-analysis. However, the resolution of the method is insufficient to resolve between the cytoplasm proper and the contents of the endoplasmatic reticulum, so that the sodium could well have been localized in the latter. This would be consistent with the analysis of, for instance SALTMAN et al. (1963), showing more sodium in the cell fluid than in the cytoplasmic fraction.

Turning back to the question of how to produce correlations between ATPases and ion transport, the similarity in sensitivity towards the cardiac glycoside ouabain has been one of the corner-stones for the parallelizations between $(Na^+ + K^+)$-activation of the biochemical system and sodium extrusion in animals. So far, ouabain has not been found very active in the biochemical systems of plants, although there have been reports on activity *in vivo* (review by KYLIN, 1973). The Russian reports by BOWLING et al. (1972) and by KRASAVINA and VYSKREBENTSEVA (1972) open new possibilities in this respect. They noted inhibition of the ATPase system by ouabain. Furthermore, they found that transport of sodium and potassium *in vivo* was inhibited when the plants were grown so that translocation was active and took place against an electrochemical gradient; but no inhibition was obtained when the electropotentials indicated that transport was mainly passive.

A correlation between transport and ATPase activity in cotyledons of *Phaseolus vulgaris* was demonstrated by LAI and THOMPSON (1972a). Membrane vesicles, that contained ATPase activated by sodium or potassium, were able to extrude the monovalent ions when incubated with ATP. The activity changed with the age of the cotyledons (LAI and THOMPSON, 1972b).

Summarizing at this point, it seems reasonable that systems for active transport of sodium, biochemically traceable as $(Na^+ + K^+)$-activated ATPases, play a role as one of the cell functions behind salt tolerance. They were first traced by an ecological approach. When put in conjunction with the ionic relations of non-vacuolized and vacuolized cells, the latter represented by the properties of inbred strains of sugar beet, the data lead to the same conclusions as electrochemical measurements, namely that there are different transport systems in the plasmalemma and in the tonoplast, working in opposite directions and keeping the cytoplasm as low in sodium as possible. It has also been possible to produce correlations with ouabain effects and vesicular transport. The main weakness of the hypothesis in its present state is that there are so many different plant species involved. It would be desirable to collect all types of data from one single system. Besides the philosophical weakness inherent in the situation, it is difficult to estimate which of the possible discrepancies are due to genetic variation between species and which depend on phenotypical adaptation.

d) Other Ion-Specific ATPases. In many cases, chloride is one of the main problems for a halophyte. A combined ecological and biochemical approach was used to trace a Cl^--activated ATPase in the salt gland of *Limonium* (HILL and HILL, 1973). The activity of the salt gland can be induced by 0.1 M NaCl. A roughly microsomal homogenate of the salt gland contains ATPase activity, which is higher in the presence of chloride than when sulphate is the only anion of the assay system. The capacity for stimulation of the ATPase by chloride is induced in the same time span as the activity of the intact gland; and both the activity of the intact gland and the development of the Cl^--stimulated ATPase are inhibited by d-actinomycin and by puromycin during the induction period but not by later applications.

With regard to the correlation between ATPase activity and ion transport, a highly critical work started from the observation of ATPases stimulated by monovalent cations in oat roots (FISHER and HODGES, 1969). As the next step, it was shown that the Rb^+-stimulated ATPase in roots of corn, wheat and barley is quantitatively sufficient to account for the transport of rubidium (FISHER et al., 1970). After working out methods to differentiate between membranes of different origins, one of the main K-activated ATPases in oat roots could be localized in the plasmalemma (LEONARD et al., 1973). Kinetic comparisons between the transport of potassium and the K^+-stimulated ATPase of the plasmalemma fraction coincide closely (LEONARD and HODGES, 1973), so closely in fact that they can be interpreted according to the multiphasic model for ion transport (NISSEN, 1973a, b; oral communication). An ATPase stimulated by monovalent cations has also been assigned to provacuolar vesicles and to the tonoplast of corn roots (LEIGH et al., 1973). The cited localization of cation-stimulated ATPases in the tonoplast as well as in the plasmalemma agrees with the situation indicated by KYLIN and HANSSON (1971) for the system activated by $(Na^+ + K^+)$.

The work by FISHER et al. (1970) has been criticized by RATNER and JACOBY (1973). The ground for the criticism was that the ATPase could be stimulated not only by inorganic, monovalent cations, but also by tris, choline, and ethanolamine. Since the inorganic cations were translocated and the organic ones not, the ATPase was regarded as unspecific. However, it is well known that tris interferes with the transport system for potassium (VAN STEVENINCK, 1962). The data of RATNER and JACOBY (1973) may then be interpreted in the sense that the organic cations can activate the "ATPase site" although they are not translocated by the "carrier site" of a transport system according to PARDEE (1968; cf. Section C. II. 1.a above).

e) Effects of Different Salt Environments. Changes in the pattern of ion transport have been demonstrated in several papers. PITMAN et al. (1968) showed that barley roots of low salt status take up monovalent cations indiscriminately or with a preference for sodium, in which case protons are given off in exchange. Roots of high salt status show a potassium-for-sodium exchange—a "sodium outpump"—which is modified by the supply of oxygen (PITMAN, 1969). POOLE (1971a, b) reported that discs of red beet show no uptake of sodium when washed for one day only, but accumulate sodium in preference to potassium after extensive washing for 6 days.

Adaptation to changes in the salt environment can occur within hours in *Zea mays*, and is dependent upon aerobic metabolism although no increase in respiration is observed (LEONARD and HANSON, 1972a). Washing was again accompanied by an increased accumulation of ions, which could be prevented by inhibitors of the synthesis of RNA and protein. Auxins, kinetin and 2,4-D also prevented the change in ion uptake. Parallel changes were noted for K^+-stimulated ATPase (LEONARD and HANSON, 1972b), giving one more correlation between ion transport and ATPase activity. The parallelisms between the levels of ATPases stimulated by Ca^{2+} and by Mg^{2+}, and the edaphic demands of wheat and oat have already been noted in Subchapter (B), as well as their response to the salt level of the medium (KYLIN and KÄHR, 1973 and unpublished).

Evidently, there is ample reason to think of responses in the cell membranes, reflected in ion transport and biochemistry, as parts of the functional system that determines whether a plant behaves as a halophyte or as a glycophyte.

2. *Membrane Structure*

There is direct structural evidence that membranes are affected by changes in the ionic environment, even though the causal relationships cannot yet be generalized. JACKMAN and VAN STEVENINCK (1967) and VAN STEVENINCK and JACKMAN (1967) observed that the changes in ion absorption capacity that are induced in beet tissue upon slicing are paralleled by changes in the membranes of mitochondria and of the endoplasmatic reticulum. Both sulphate and chloride salinization caused changes in the membranes of *Atriplex halimus* (BLUMENTHAL-GOLDSCHMIDT and POLJAKOFF-MAYBER, 1968). Conspicuous differences were observed in membranes of *Zea mays*, dependent upon whether the tissues were depleted of potassium or treated with sodium (HECHT-BUCHHOLZ and MARSCHNER, 1970; HECHT-BUCHHOLZ, 1971; HECHT-BUCHHOLZ et al., 1971). UDOVENKO et al. (1970) applied NaCl and polyethyleneglycol to three bean varieties of different salt resistance. The changes induced were of the same type in all varieties, but the time for recovery was shorter for resistant than non-resistant material. Furthermore, the sequence from breakdown to normalization was quicker in the rapidly metabolizing mitochondria and endoplasmatic reticulum than in the leucoplasts. A sequence of breakdown and restoration of membranes upon osmotic shock and subsequent rehydration was reported also by NIR et al. (1969, 1970). Disorganization of the relationship between proteins and lipids in the membranes may have occurred, since oxygen consumption decreased at the same time as cytochrome oxidase activity in the mitochondria increased (NIR et al., 1970a). Histochemical methods have disclosed the presence of ATPases in the membranes (MCCLURKIN and MCCLURKIN, 1967; HALL, 1969).

3. Chemical Changes

Relations between salt transport, salt tolerance and lipids were observed by KUIPER (1968a, b). Using a series of five rootstocks of grape varieties of different salt tolerance, he found that their content of monogalactosyl diglyceride increased with increasing capacity for uptake of chloride and salt sensitivity. Phosphatidyl choline and phosphatidyl ethanolamine were correlated with the same functions in the reverse manner, and were particularly rich in lignoceric acid (saturated C_{24} in a straight chain). In model experiments monogalactosyl diglyceride facilitated transport of chloride through a layer of pentanol between two water phases, whereas phosphatidyl choline facilitated exchange of sodium for potassium. In later experiments with bean and cotton (KUIPER, 1969), addition of different lipids increased uptake of chloride to the roots. Neutral and acidic lipids increased ion transport to the shoots, but the zwitterionic phosphatidyl choline had no effect in the aerial parts, since it was absorbed only by the root tissue. However, glycerophosphoryl choline was transported from the root medium to all plant parts and, presumably, esterified to phospholipid, after which chloride accumulation in the leaves was reduced.

The investigations were extended to the relationships between enzymatic action and lipids (KUIPER, 1972a). A defatted membrane protein without ATPase activity could be reconstituted as an ATPase with the help of phosphatidyl choline and sulfolipid. Each lipid gave about 40% reconstitution, but the temperature-dependence of the reconstituted ATPases differed in the two cases.

Against the background of the reconstitution data, it was decided to analyze how the lipid composition varies when intact roots are compared with ATPases from differently treated microsomal preparations. The ATPase fraction was investigated as an untreated "control" as well as after treatments with desoxycholate, cysteine or dithiothreitol, which develop the specificity for $(Na^+ + K^+)$ to different degrees and give different specific activities to the preparations (KYLIN et al., 1972; HANSSON et al., 1973). High retention of the acidic sulfolipid within the membrane fragments and high loss of the zwitterionic phosphatidyl choline are correlated with high specificity and activity of the $(Na^+ + K^+)$-stimulated ATPase.

The interpretation of the cited experiments on reconstitution and analysis appears to be that different types of ATPase are related to the lipid composition of the membrane. One must assume that *in vivo* different lipids can be brought to cooperate with the membrane proteins so as to form active sites that respond to the changes of the environment in a suitable manner that has been determined by the selection pressure of evolution.

With regard to $(Na^+ + K^+)$-activated ATPases and related phenomena, the involvement of a highly acidic lipid has also been found in other biological systems: sulfatides in the salt gland of the duck (K.-A. KARLSSON et al., 1971) and in Halobacterium (KATES et al., 1972), phosphatidyl serine in erythrocytes (ROELOFSEN, 1968). A high negativity around the active site is needed to bind the very mobile monovalent cations to the structure, and the neutralization of the negativity by the cations may be necessary for the ATPase function (KYLIN et al., 1972). Measurements of the surface charge of the particles from the roots of the sugar beet seedlings show an influence of the preparation method (J. KARLSSON et al., 1971) as well as of the ionic medium (J. KARLSSON and TRIBUKAIT, unpublished).

Correlations between the salinity of irrigation water and the lipids of cotton have been found by TWERSKY and FELHENDLER (1973). In the roots of wheat and oat, some lipids vary directly with the salinity of the medium, whereas others follow the adaptative level of the ATPase as described in Subchapter (B). The following is a summary of ATPase data by KYLIN and KÄHR (1973 and unpublished) and lipid analysis by KUIPER and STUYVER (unpublished):

Table 1. Effect of high salt condition

Property or constituent analyzed	Oat	Wheat	Lipid fraction follows variation in
ATPase activity	Decrease	Increase	—
Sterols (glucosides)	Decrease	Decrease	Salinity
Glycolipids	Decrease	Decrease	Salinity
(Phosphatidyl) choline	Decrease	Increase	ATPase
Total phospholipids	Decrease	Increase	ATPase
Sulfolipids	Decrease	Increase	ATPase

As a matter of course, the behavior of membranes is a matter of proteins as much as of lipids. However, no directly relevant data seem to be available. Possibly, one may associate the high contents of proline that have been reported in connection with wilting (STEWART, 1972 for references) and with salt stress (GOAS, 1965; 1967), with the finding by KUIPER and LIVNE (1973) that the membranes of the camel erythrocyte contain an unusually high proportion of protein. This protein is high in proline and arginine, which indicates a high lipid:protein interaction and a low degree of formation of helices in the protein (l.c.).

III. Nucleic Acids and Protein Metabolism

Even though changes in the outer cell membranes, in direct response to changes in the ionic environment and to ion transport, may be thought of as the first part of adaptation to saline environments, it intuitively seems that long-term adjustment should be linked to the metabolism of nucleic acids and the ensuing control over the synthesis of enzymatic proteins. In the present chapter this aspect will be dealt with only cursorily, since the field is so closely related to that of hormones and growth that it may better belong elsewhere.

The balance between soluble amino acids and protein is changed by salination (STROGONOV, 1964). KAHANE and POLJAKOFF-MAYBER (1968) studied the uptake and incorporation of labelled amino acids into pea roots. Both functions were reduced by the presence of salt in the incubation medium. If the roots were pregrown with salt and the experiments performed without salt stress, uptake and incorporation were still reduced as compared to the controls, indicating that a more permanent effect had been induced. Sulphate salinity was more deleterious than salinity due to chloride. In the interaction with kinetin, non-stressed and sulphate-stressed roots responded with decreased incorporation, chloride-stressed roots with an increase.

The changes in the balance between proteins and amino acids induced by salinity is reminiscent of the events occurring during drought (VAADIA et al., 1961). Also the interaction between kinetin and the incorporation of amino acids is found during osmotic stress without relation to ions (BEN-ZIONI et al., 1967). One finds a breakdown of polysomes during stress (NIR et al., 1970b), which can be accompanied by an increased level of RNAase (MARIN and VIEIRA DA SILVA, 1972). Drought resistant species show less breakdown of the polysomes than drought sensitive, but there is no relationship between polysomes and RNAase activity in the drought resistant *Robinia pseudacacia* (BRANDLE et al., 1973).

The whole sequence DNA-RNA-protein has been investigated in extensive comparisons between the salt sensitive *Citrus limettoides Tanaka* (sweet lime) and the salt resistant C. *reticulata Blance* (Cleopatra mandarin). The general picture appears to imply that high salt conditions increase the breakdown due to proteolytic and nucleolytic processes, but in the salt resistant variety this is more than compensated for by increased rates of synthesis triggered by the salt (KESSLER, 1961, 1966; KESSLER and CHEN, 1964; KESSLER and SNIR, 1969). It was assumed that the intensified synthesis of protein under the action of salt is induced by the removal of basic proteins from the DNA of the salt tolerant plants (KESSLER, 1966). Biochemical evidence is cited, to the effect that changes in the balance between potassium and ammonium versus sodium and lithium can change the polymerization processes that give rise to new nucleic acids (LUBIN, 1963; CONWAY and LIPMAN, 1964). Thus the possibility would be open to translate changes in the ionic environment via membrane transport to changes in the central control mechanism of the cell.

IV. General Metabolism

Several sequences of general metabolism have been analyzed in their dependance on salinity. JOSHI et al. (1962) showed that marine plants produce much more amino acids and less organic acids than terrestrial plants, when they fix carbon dioxide in the dark. In a homogenate from spinach leaf, the per cent carbon dioxide fixed as amino acids was greater in the presence of NaCl than in its absence. Likewise, obligate salt marsh plants produce comparatively more amino acids than non-obligate ones (WEBB and BURLEY, 1965); and the non-obligate *Spartina alterniflora* is pushed over to the production of amino acids, when it is treated with sodium chloride (l.c.). Corresponding evidence for photosynthesizing spinach chloroplasts was given by GEE et al. (1965).

Combining the above data with the observations by EVANS and SORGER (1966) that monovalent ions influence various enzyme activities, WEIMBERG (1967) formed the hypothesis that the change from production of organic acids to production of amino acids may be due to the interaction of sodium with malic dehydrogenase. He showed that the activity of the enzyme will increase up to a concentration of 0.02 M NaCl, whereas higher concentrations of the salt are supraoptimal. The maximal velocity (V_{max}) of the reaction will increase, but in the range of 0.02–0.1 M NaCl the salt will competitively decrease the affinity of the enzyme for oxaloacetate without affecting the affinity for NADH. The biochemical data coincide well with the physiological reality, since they indicate that the interaction of salt with malic dehydrogenase can lead to an increased production of oxaloacetate for further transformation to aspartate. The ensuing high demand for reduced nitrogen can be correlated with a high activity of nitrate reductase in halophytes (BOUCAUD, 1972).

A long series of work from the group of POLJAKOFF-MAYBER has already been cited (Section B), demonstrating how differently chloride salinity and sulphate salinity affect the balance between the pentose phosphate pathway, the Krebs cycle, and so on. Parallel to the works cited on the relationship between salt status and ion transport and ATPases (Section C.II.1.e.), one may mention the paper by LIVNE and LEVIN (1967). They found that pea roots grown in a saline medium had higher respiration rates than control roots grown in a normal nutrient. Furthermore, since the mitochondria of the high-salt roots were well coupled, there was a higher rate of oxidative phosphorylation. The mitochondria from control roots were inhibited by addition of salt to the preparation medium, the mitochondria from salt treated roots were not influenced by 75 mM NaCl. Evidently, adaptative changes had taken place.

Competition between phosphoenolpyruvate carboxylase and ribulosediphosphate carboxylase may be decisive for whether the photosynthesis of a plant follows the C_4 or the C_3 pathway (COOMBS, 1971). As mentioned, SHOMER-ILAN and WAISEL (1973) investigated how the activities of the two enzymes are influenced by the addition, for several days, of 100 mM NaCl to a half-strength Hoagland solution. Both enzymes were on the whole adversely affected in the salt sensitive *Zea mays*. In the medium salt tolerant *Chloris gayana* one gets the impression that the activities first decline and then recover. Both these species are C_4 plants. In the halophytic *Aeluropus litoralis* both enzymes increase in activity with the length of the salt application and with the age of the leaf tested. During this increase, the ratio of the two enzymes is changed. At the start, the phosphoenolpyruvate carboxylase is about one tenth or less of the ribulosediphosphate carboxylase. At the end of the treatment the proportion is about one to three. Concomitantly, the plants change from C_3 to C_4 assimilation of CO_2, providing experimental evidence for the idea of COOMBS (1971). It should be noted that this effect of sodium is hardly related to the role of Na as a trace element for C_4 plants (BROWNELL and CROSSLAND, 1972). The role as trace element is filled by additions of 0.1 mM, which is 1000 times less than the concentration used by SHOMER-ILAN and WAISEL (1973). One is rather reminded of an earlier investigation by BROWNELL (1968), pointing out that the amounts of Na that favor *Beta* and *Chenopodium* indicate that these plants have another mechanism of action than *Atriplex*, where sodium is a trace element.

Respiration of exogenous substrate is much more inhibited than endogenous respiration if the water potential of the medium is lowered. This has been observed in *Chlorella* (GREENWAY and HILLER, 1967; HILLER and GREENWAY, 1968) as well as for different yeasts (NORKRANS, 1968). For the yeasts, the lowest tolerated water potentials were obtained with glycerol as osmoticum, various sugars were intermediate, and salts of Na and K the most deleterious. Halotolerance was coupled to the ability to maintain energy metabolism at high levels of salt—up to 16 or 20% NaCl for the most extreme species; and to the utilization of the energy to extrude sodium in exchange for potassium, whereas the salt sensitive baker's yeast acidified the medium instead of sending out Na (NORKRANS and KYLIN, 1969).

D. Conclusions

In a way, the last reference closes a circle. Under changing salt conditions in the medium, the transport processes in the plant membranes are modified, be it active transport mediated by ATPases or passive translocation. Changes in the internal ionic composition can lead to adjustments of inner membranes, and perhaps also to responses in the nucleic acids themselves, the central system for long-time regulation. The regulatory influence will sooner or later change the amounts and connections of energy metabolism, which in turn will affect ion transport. The ideas can be summarized in the scheme below, where the words are intended to give reference to various subchapters, and the arrows intended to show possible interconnections.

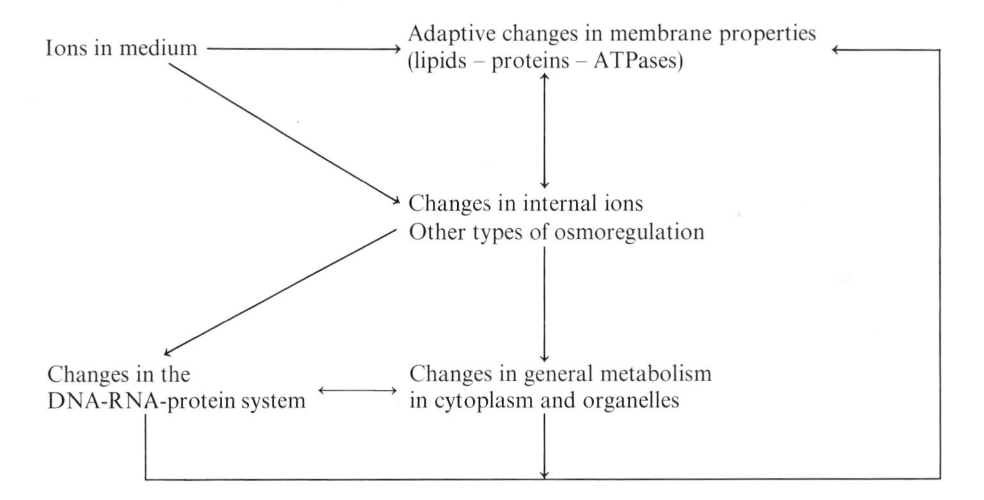

The picture is oversimplified—the physiological and biochemical processes underlying halotolerance are certainly more diversified than the circle drawn. At the same time the picture may be too complex: it is often advisable in the planning of experiments to regard the cell functions involved as independent parts and not give too much thought to the whole. Nevertheless, sooner or later one must start to worry about a theoretical system that is complex enough to allow several functions to balance each other through regulation and feedback, so that an ecologically induced response can be described. It is the belief of the authors that a start can be made by using specializations such as halotolerant plants for investigations where ecology and biochemistry are allowed to meet.

Acknowledgements

For the senior author, the present paper is the last one from an institute, to which he has been affiliated for twelve years and in which he has reached what may hopefully be regarded as scientific maturity. The years can be characterized by the words friendship and trust; and I wish to express my gratitude to all my colleagues by directing my thanks to Professor TORSTEN HEMBERG of the University of Stockholm, Sweden.

The review was initiated during a visit of the senior author to the Botany Department of the Oregon State University in Corvallis, Oregon. A great part of the references was collected during a stay in the Botany Department of the University of Tel Aviv in Israel, despite the energy with which Professor YOAV WAISEL tried to teach me practical rather than theoretical things about halophytes.

References

AHMAD, R., HEWITT, E.J.: Studies on the growth and phosphatase activities in *Suaeda fruticosa*. Plant and Soil **34**, 691–696 (1971).

BAXTER, R.M., GIBBONS, N.E.: The glycerol dehydrogenases of *Pseudomonas salinasia*, *Vibrio costicolus*, and *E. coli* in relation to bacterial halophilism. Can. J. Biochem. Physiol. **32**, 206–217 (1954).

BEN-AMOTZ, A., AVRON, M.: Photosynthetic activities of the halophilic alga *Dunaliella parva*. Plant Physiol. **49**, 240–243 (1972).

BEN-AMOTZ, A., AVRON, M.: The role of glycerol in the osmotic regulation of the halophilic alga *Dunaliella parva*. Plant Physiol. **51**, 875–873 (1973).

BENSON, A.A.: Plant membrane lipids. Ann. Rev. Pl. Physiol. **15**, 1–16 (1964).

BEN-ZIONI, A., ITAI, C., VAADIA, Y.: Water and salt stresses, kinetin and protein synthesis in tobacco leaves. Plant Physiol. **42**, 361–365 (1967).

BLUMENTHAL-GOLDSCHMIDT, S., POLJAKOFF-MAYBER, A.: Effect of substrate salinity on growth and on submicroscopic structure of leaf cells of *Atriplex halimus* L. Australian J. Botan. **16**, 469–478 (1968).

BOUCAUD, J.: Effets du NaCl sur l'activité du nitrate réductase chez deux variétées halophiles: *Suaeda maritima* (L.) Dum. variété *macrocarpa* (MOQ) et variété *flexilis* (FOCKE). Physiol.Plant. **27**, 37–42 (1972).

BOUILLENNE-WALRAND, M., BOUILLENNE-WALRAND, R.: Contribution á l'étude de la respiration en fonction de l'hydratation. Échanges respiratoires dans les racines tuberisées de *Brassica napus* L. Ann. Physiol. Physiochim. Biol. **2**, 426–467 (1926).

BOWLING, I.F., TURKINA, M.V., KRASAVINA, M.S., KRYUCHESNIKOVA, A.L.: Na^+–, K^+-activated ATPase of conducting tissues. Fiziol. Rast. (Transl.) **19**, 824–832 (1972).

BRANDLE, I.R., SCHNARE, P.D., HINCKLEY, T.M., BROWN, G.N.: Changes in polysomes of black locust seedling during dehydration-rehydration cycles. Physiol. Plant. **29**, 406–409 (1973).

BROWNELL, P.F.: Sodium as an essential micronutrient for some higher plants. Plant and Soil **28**, 161–164 (1968).

BROWNELL, P.F., CROSSLAND, C.J.: The requirement for sodium as a micronutrient by species having the C_4 dicarboxylic photosynthetic pathway. Plant Physiol. **49**, 794–797 (1972).

COLLANDER, R.: Selective absorption of cations by higher plants. Plant Physiol. **16**, 691–720 (1941).

CONWAY, T.W., LIPMAN, F.: Characterization of a ribosome-linked guanosine triphosphatase in *Escherichia coli* extracts. Proc. Natl. Acad. Sci. U.S. **52**, 1462–1469 (1964).

COOMBS, J.: The potential of higher plants with the phosphopyruvic acid cycle. Proc. Roy. Soc. London B **179**, 221–235 (1971).

DAINTY, J.: Ion transport and electrical potentials in plant cells. Ann. Rev. Plant Physiol. **13**, 379–402 (1962).

DAUBENMIRE, R. F.: Plants and Environment. 2nd ed. New York: Wiley 1959.

EISENMAN, G.: On the elementary atomic origin of equilibrium ionic specificity. KLEINZELLER, A., and KOTYK, A. (Eds.): Membrane transport and metabolism, pp. 163–179. New York: Academic Press 1960.

EL-SHEIKH, A. M., UHLRICH, A., BROYER, T. C.: Sodium and rubidium as possible nutrients for sugar beet plants. Plant. Physiol. **42**, 1202–1208 (1967).

EVANS, H. J., SORGER, G. J.: Role of mineral elements with emphasis on the univalent cations. Ann. Rev. Plant. Physiol, **17**, 47–76 (1966).

FISHER, J., HANSEN, D., HODGES, T. K.: Correlation between ion fluxes and ion-stimulated adenosine triphosphatase activity of plant roots. Plant Physiol. **46**, 812–814 (1970).

FISHER, J., HODGES, T. K.: Monovalent ion stimulated adenosine triphosphatase from oat roots. Plant Physiol. **44**, 385–395 (1969).

FLOWERS, T. I.: Salt tolerance in *Suaeda maritima* (L.) Dum. The effect of sodium chloride on growth, respiration, and soluble enzymes in a comparative study with *Pisum sativum* L. J. Exp. Botan. **23**, 310–321 (1972a).

FLOWERS, T. I.: The effect of sodium chloride on enzyme activities from four halophyte species of *Chenopodiaceae*. Phytochemistry **11**, 1881–1886 (1972b).

GEE, R., JOSHI, G., BILS, R. F., SALTMAN, P.: Light and dark CO_2 fixation by spinach leaf systems. Plant Physiol. **40**, 89–96 (1965).

GINZBURG, M.: The unusual membrane permeability of two halophilic unicellular organisms. Biochim. Biophys. Acta **173**, 370–376 (1969).

GOAS, M.: Sur le métabolisme azote des halophytes: Études des acides amines et amides libres. Soc. Franc. Physiol. Véget. Bull. **11**, 309–316 (1965).

GOAS, M.: Contribution á l'étude du métabolisme azote des halophytes: Acides amines et amides libres d'*Aster tripolium L.* en aquiculture. C.R. Séances Acad. Sci. (Paris) Ser. D **265**, 1049–1052 (1967).

GRAZIANI, Y., LIVNE, A.: Water permeability of bilayer lipid membranes: sterol-lipid interaction. J. Membrane Biol. **7**, 275–284 (1972).

GREEN, D. E., TZAGOLOFF, A.: Role of lipids in the structure and function of biological membranes. J. Lipid Res. **7**, 587–602 (1966).

GREENWAY, H.: Effects of slowly permeating osmotica on metabolism of vacuolated and nonvacuolated tissues. Plant Physiol. **46**, 254–258 (1970).

GREENWAY, H., HILLER, R. G.: Effects of low water potentials on respiration and on glucose and acetate uptake by *Chlorella pyrenoidosa*. Planta **75**, 253–274 (1967).

GREENWAY, H., LEAHY, M.: Effects of rapidly and slowly permeating osmotica on metabolism. Plant Physiol. **46**, 259–262 (1970).

GREENWAY, H., OSMOND, C. B.: Salt responses of enzymes from species differing in salt tolerance. Plant Physiol. **49**, 256–259 (1972).

HACKETT, D. P.: Effects of salts on DPNH oxidase activity and structure of sweet potato mitochondria. Plant Physiol. **36**, 445–452 (1961).

HALL, J. L.: A histochemical study of adenosine triphosphatase and other nucleotide phosphatases in young root tips. Planta **89**, 254–265 (1969).

HALL, J. L.: Further properties of adenosine triphosphatase and β-glycerophosphatase from barley roots. J. Exp. Botan. **22**, 800–808 (1971).

HANSSON, G., KUIPER, P. J. C., KYLIN, A.: Effect of preparation method on the induction of (sodium + potassium)-activated adenosine triphosphatase from sugar beet root and its lipid composition. Physiol. Plant. **28**, 430–435 (1973).

HANSSON, G., KYLIN, A.: ATPase activities in homogenates from sugar beet roots, relation to Mg^{2+} and $(Na^+ + K^+)$-stimulation. Z. Pflanzenphysiol. **60**, 270–275 (1969).

HASSON-PORATH, E., POLJAKOFF-MAYBER, A.: The effect of salinity on the malic dehydrogenase of pea roots. Plant Physiol. **44**, 1031–1034 (1969).

HASSON-PORATH, E., POLJAKOFF-MAYBER, A.: Lactic acid content and formation in pea roots exposed to salinity. Plant Cell Physiol. **11**, 891–897 (1970).

HASSON-PORATH, E., POLJAKOFF-MAYBER, A.: Content of adenosine phosphate compounds in pea roots grown in saline media. Plant Physiol. **47**, 109–113 (1971).

HECHT-BUCHHOLZ, CH.: The effect of potassium deficiency on fine structure of proplastids. Skokloster colloquium on potassium in biochemistry and physiology, pp. 40–49. Bern: Intern. Potash Inst. 1971 (1973).

HECHT-BUCHHOLZ, CH., MARSCHNER, H.: Veränderungen der Feinstruktur von Zellen der Maiswurzelspitze bei Entzug von Kalium. Z. Pflanzenphysiol. **63**, 416–427 (1970).

HECHT-BUCHHOLZ, CH., PFLÜGER, R., MARSCHNER, H.: Einfluß von Natriumchlorid auf Mitochondrienzahl und Atmung von Maiswurzelspitzen. Z. Pflanzenphysiol. **65**, 410–417 (1971).

HEXUM, T., SAMSON, E., HIMES, R.: Kinetic studies of membrane ($Na^+ - K^+ - Mg^{2+}$)-ATPase. Biochim. Biophys. Acta **212**, 322–331 (1970).

HILL, B.S., HILL, A.K.: Enzymatic approaches to chloride transport in the *Limonium* salt gland, pp. 379–384. In: ANDERSON, W.P. (Ed.). Ion transport in plants. London, New York: Academic Press (1972) 1973.

HILLER, R.G., GREENWAY, H.: Effects of low water potentials on some aspects of carbohydrate metabolism in *Chlorella pyrenoidosa*. Planta **78**, 49–59 (1968).

HIPPEL, P.H., VON, SCHEICH, T.: The effects of neutral salts on the structure and conformational stability of macromolecules in solution, pp. 417–574. In: TIMASHEFF, S.W., FASSMAN, G.D. (Eds.): Structure and stability of biological macromolecules. New York: Marcel Decker Inc. 1969.

HOPFER, U., LEHNINGER, A.L., LENNARZ, W.J.: The effect of the polar moiety of lipids on bilayer conductance induced by uncouplers of oxidative phosphorylation. J. Membrane Biol. **3**, 142–155 (1970).

JACKMAN, M.E., VAN STEVENINCK, R.F.M.: Changes in the endoplasmic reticulum of beet root slices during aging. Australian J. Biol. Sci. **20**, 1063–1068 (1967).

JOHNSON, M.K., JOHNSON, R.D., MACELROY, R.D., SPEER, H., BRUFF, B.S.: Effects of salts on the halophilic alga *Dunaliella viridis*. J. Bacteriol. **95**, 1461–1468 (1968).

JOKELA, A.C.-C.T.: Outer membrane of *Dunaliella tertiolecta*: Isolation and properties. Ph. D. Dissertation, Univ. of Calif. San Diego (1969).

JOSHI, G., DOLAN, T., GEE, R., SALTMAN, P.: Sodium chloride effect on dark fixation of CO_2 by marine and terrestrial plants. Plant Physiol. **37**, 446–449 (1962).

KAHANE, I., POLJAKOFF-MAYBER, A.: Effect of substrate salinity on the ability for protein synthesis in pea roots. Plant Physiol. **43**, 1115–1119 (1968).

KARLSSON, J., TRIBUKAIT, B., KYLIN, A.: Electrical mobilities of fragments of sugar beet root membranes with properties as ($Na^+ + K^+$)-activated adenosine triphosphatases. Proc. 1st. European Biophys. Congr. **III**, 75–79 (1971).

KARLSSON, K.-A., SAMUELSSON, B.E., STEEN, G.O.: Lipid pattern and $Na^+ - K^+ -$dependent adenosine triphosphatase activity in the salt gland of duck before and after adaptation to hypertonic saline. J. Membrane Biol. **5**, 169–184 (1971).

KATES, M., HANCOCK, A.J., DEROD, P.W.: Phosphatide and glycolipid sulfate esters in cell envelope of *Halobacterium cutirubrum*. Abst. Commun. Meet. Fed. European. Biochem. Soc. **8**, 124 (1972).

KAUSS, H.: Metabolism of isofloridoside and osmotic balance in the fresh water alga *Ochromonas*. Nature **214**, 1129–1130 (1967).

KAVANAU, J.L.: Water and solute-water interactions. San Fransisco: Holden-Day 1964.

KESSLER, B.: Nucleic acids as factors in drought resistance. Recent Advanc. Botan. **2**, 1153–1159 (1961).

KESSLER, B.: The physiological basis of the tolerance of evergreen trees to lime and saline soil and water conditions, with special reference to the selection of root stocks of Avocado and Citrus by physiological tests. Final report. Rehovot, Israel: Volcani Inst. of Agric. Res. Nov. 1966.

KESSLER, B., CHEN, D.: Mediumspecific activity of polynucleotide phosphorylase. 6th Int. Biochem. Congr. (Moscow) **I**, 65 (1964).

KESSLER, B., SNIR, I.: Salt effects on nucleic acids and protein metabolism in Citrus seedlings. Proc. 1st. Int. Citrus Symp. **1**, 381–386 (1969).

KLIS, F.M.: α-Glucosidase activity at the cell surface in callus of *Convolvolus arvensis*. Physiol. Plant. **25**, 253–257 (1971).

KLOTZ, I.M.: Protein hydration and behavior. Science **128**, 815–822 (1958).

KRASAVINA, M.S., VYSKREBENTSEVA, E.I.: ATPase activity and transport of potassium and sodium in root tissues. Fiziol. Rast. (Transl.) **19**, 833–837 (1972).

KUIPER, P.J.C.: Lipids in grape roots in relation to chloride transport. Plant Physiol. **43**, 1367–1371 (1968a).

KUIPER, P.J.C.: Ion transport chracteristics of grape root lipids in relation to chloride transport. Plant Physiol. **43**, 1372–1374 (1968b).

KUIPER, P.J.C.: Effect of lipids on chloride and sodium transport in bean and cotton plants. Plant Physiol. **44**, 968–972 (1969).

KUIPER, P.J.C.: Temperature response of adenosine triphosphatase of bean roots as related to growth temperature and to the lipid requirement of the adenosine triphosphatase. Physiol. Plant. **26**, 200–205 (1972a).

KUIPER, P.J.C.: Water transport across membranes. Ann. Rev. Plant Physiol. **23**, 157–172 (1972b).

KUIPER, P.J.C., LIVNE, A.: Properties of the camel erythrocyte membrane. Abstr. 9th. Intern. Congr. Biochem, p. 281. Stockholm 1973.

KYLIN, A.: An outpump balancing phosphate-dependent sodium uptake in *Scenedesmus*. Biochem. Biophys. Res. Commun. **16**, 497–500 (1964).

KYLIN, A.: Uptake and loss of Na^+, Rb^+, and Cs^+ in relation to an active mechanism for extrusion of Na^+ in *Scenedesmus*. Plant Physiol. **41**, 579–584 (1966).

KYLIN, A.: Adenosine triphosphatases stimulated by (sodium + potassium): Biochemistry and possible significance for salt resistance, pp. 369–377. In: ANDERSON, W.P. (Ed.): Ion Transport in Plants. London, New York: Academic Press (1972) 1973.

KYLIN, A., GEE, R.: Adenosine triphosphatase activities in leaves of the mangrove *Avicennia nitida* Jacq. Influence of sodium to potassium ratios and salt concentrations. Plant Physiol. **45**, 169–172 (1970).

KYLIN, A., HANSSON, G.: Transport of sodium and potassium, and properties of (sodium + potassium)-activated adenosine triphosphatases: possible connection with salt tolerance in plants. Skokloster colloquium on potassium in biochemistry and physiology, pp. 64–68. Bern: Intern. Potash Inst. 1971 (1973).

KYLIN, A., KÄHR, M. The effect of magnesium and calcium ions on adenosine triphosphatases from wheat and oat roots at different pH. Physiol. Plant. **28**, 452–457 (1973).

KYLIN, A., KUIPER, P.J.C., HANSSON, G.: Lipids from sugar beet in relation to the preparation and properties of (sodium + potassium)-activated adenosine triphosphatases. Physiol. Plant. **26**, 271–278 (1972).

LAI, Y.F., THOMPSON, J.E.: The preparation and properties of an isolated plant membrane fraction enriched in $(Na^+ + K^+)$ stimulated ATPase. Biochim. Biophys. Acta **233**, 84–90 (1971).

LAI, Y.F., THOMPSON, J.E.: Effect of germination on $(Na^+ - K^+)$-stimulated adenosine 5′-triphosphatase and ATP-dependent ion transport of isolated membranes from cotyledons. Plant Physiol. **50**, 452–457 (1972a).

LAI, Y.F., THOMPSON, J.E.: The behavior of ATPase in cotyledon tissue during germination. Can. J. Bot. **50**, 327–332 (1972b).

LATIES, G.G.: The osmotic inactivation *in situ* of plant mitochondrial enzymes. J. Exp. Botan. **5**, 49–70 (1954).

LEIGH, R.A., WYN JONES, R.G., WILLIAMSON, F.A.: The possible role of vesicles and ATPases in ion uptake, pp 407–418. In: ANDERSON, W.P. (Ed.): Ion transport in plants. London, New York: Academic Press (1972) 1973.

LEONARD, R.T., HANSEN, D., HODGES, T.K.: Membrane-bound adenosine triphosphatase activities of oat roots. Plant Physiol. **51**, 749–754 (1973).

LEONARD, R.T., HANSON, J.B.: Induction and development of increased ion absorption in corn root tissue. Plant Physiol. **49**, 430–435 (1972a).

LEONARD, R.T., HANSON, J.B.: Increased membrane-bound adenosine triphosphatase activity accompanying development of enhanced solute uptake in washed corn root tissue. Plant Physiol. **49**, 436–440 (1972b).

LEONARD, R.T., HODGES, T.K.: Kinetics of KCl stimulated plasma membranes from oat roots. Plant Physiol. **51** (Suppl.), 234 (1973).

LIVNE, A., LEVIN, N.: Tissue respiration and mitochondrial oxidative phosphorylation of NaCl-treated pea seedlings. Plant Physiol. **42**, 407–414 (1967).

LUBIN, M.: A priming reaction in protein synthesis. Biochim. Biophys. Acta **72**, 345–348 (1963).

MARIN, B., VIEIRA DA SILVA, J.: Influence de la carence hydrique sur la repartition cellulaire de l'acide ribonucleique foliaire chez le cotonnier. Physiol. Plant. **27**, 150–155 (1972).

MAZUR, P.: Freezing injury in plants. Ann. Rev. Pl. Physiol. **20**, 419–445 (1969).

MCCLURKIN, I.T., MCCLURKIN, D.C.: Cytochemical demonstration of a sodium-activated and a potassium-activated adenosine triphosphatase in loblolly pine seedlings. Plant Physiol. **42**, 1103–1110 (1967).

MOTHES, K.: Der Einfluß des Wasserzustandes auf Fermentprozesse und Stoffumsatz. RUHLAND, W. (Ed.): Encyclopedia of plant physiology **III**, 656–664. Berlin-Göttingen-Heidelberg-New York: Springer 1956.

NIEMAN, R.H.: Salt suppression of glucose and phosphate uptake by root tissue, p. 158. XI. Int. Botan. Congr. (Seattle) Abstracts 1969.

NIR, I., KLEIN, S., POLJAKOFF-MAYBER, A.: Effect of moisture stress on submicroscopic structure of maize roots. Australian J. Biol. Sci. **22**, 17–33 (1969).

NIR, I., KLEIN, S., POLJAKOFF-MAYBER, A.: Changes in fine structure of root cells from maize seedlings exposed to water stress. Australian. J. Biol. Sci. **23**, 489–491 (1970).

NIR, I., POLJAKOFF-MAYBER, A., KLEIN, S.: The effect of water stress on mitochondria of root cells. A biochemical and cytochemical study. Plant Physiol. **45**, 173–177 (1970a).

NIR, I., POLJAKOFF-MAYBER, A., KLEIN, S.: The effect of water stress on the polysome population and the ability to incorporate amino acids in maize root tips. Israel J. Botan. **19**, 451–462 (1970b).

NISSEN, P.: Kinetics of ion uptake in higher plants. Physiol. Plant. **28**, 113–120 (1973a).

NISSEN, P.: Multiphasic ion uptake in roots, pp. 539–553. In: ANDERSON, W.P. (Ed.): Ion transport in plants. London, New York: Academic Press (1972) 1973b.

NORKRANS, B.: Studies on marine occurring yeasts: Respiration, fermentation and salt tolerance. Arch. Mikrobiol. **62**, 358–372 (1968).

NORKRANS, B., KYLIN, A.: Regulation of the potassium to sodium ratio and of the osmotic potential in relation to salt tolerance in yeasts. J. Bacteriol. **100**, 836–845 (1969).

OKAMOTO, H., SUZUKI, Y.: Intracellular concentration of ions in a halophilic strain of *Chlamydomonas*. I. Concentration of Na, K, and Cl in the cell. Z. Allg. Mikrobiol. **4**, 350–357 (1964).

ONISHI, H.: Studies on osmophilic yeasts. Part XV. The effects of high concentrations of chloride on polyalcohol production. Agr. Biol. Chem. **27**, 543–547 (1963).

OSMOND, C.B., GREENWAY, H.: Salt responses of carboxylation enzymes from species differing in salt tolerance. Plant Physiol. **49**, 260–263 (1972).

PAPAHADJOPOULOS, D.: $Na^+ - K^+$ discrimination by "pure" phospholipid membranes. Biochim. Biophys. Acta **241**, 254–259 (1971).

PARDEE, A.B.: Membrane transport proteins. Science **162**, 632–637 (1968).

PIERCE, W.S., HIGINBOTHAM, N.: Compartments and fluxes of K^+, Na^+, and Cl^- in *Avena* coleoptile cells. Plant Physiol. **46**, 666–673 (1970).

PITMAN, M.G.: Adaptation of barley roots to low oxygen supply and its relation to sodium and potassium uptake. Plant Physiol. **44**, 1233–1240 (1969).

PITMAN, M.G., COURTICE, A.L., LEE, B.: Comparison of potassium and sodium uptake by barley roots at high and low salt status. Australian. J. Biol. Sci. **21**, 871–881 (1968).

POOLE, R.J.: Effect of sodium on potassium fluxes at the cell membrane and vacuole membrane of red beet. Plant Physiol. **47**, 731–734 (1971a).

POOLE, R.J.: Development and characteristics of sodium-selective transport in red beet. Plant Physiol. **47**, 735–739 (1971b).

PORATH, E., POLJAKOFF-MAYBER, A.: Effect of salinity on metabolic pathways in pea root tips. Israel J. Botan. **13**, 115–121 (1964).

PORATH, E., POLJAKOFF-MAYBER, A.: The effect of salinity in the growth medium on carbohydrate metabolism in pea root tips. Plant Cell Physiol. **9**, 195–203 (1968).

RATNER, A., JACOBY, B.: Non-specificity of salt effects on Mg^{2+}-dependent ATPase from grass roots. J. Exp. Botan. **24**, 231–238 (1973).

ROELOFSEN, B.: Some studies on the extractability of lipids and the ATPase activity of the erythrocyte membrane. Diss. Unv. of Utrecht, Netherlands 1–66 (1968).

ROTHSTEIN, A.: Membrane function and physiological activity of microorganisms, pp. 23–39. In: HOFFMAN, J.F. (Ed.): The cellular functions of membrane transport. Englewood Cliffs, N.J.: Prentice Hall 1964.

SALTMAN, P., FORTE, J.G., FORTE, G.M.: Permeability studies on chloroplasts from *Nitella*. Exptl. Cell. Res. **29**, 504–514 (1963).

SHOMER-ILAN, A., WAISEL, Y.: The influence of sodium on the balance between the C_3- and C_4-carbon fixation pathways. Physiol. Plant **29**, 190–193 (1973).

SKOU, J.C.: Enzymatic aspects of linked transport of Na and K through the cell membrane. Progr. Biophys. Molec. Biol. **14**, 131–166 (1964).

SMITH, A.U.: Effects of low temperatures on living cells and tissues, pp. 1–62. In: HARRIS, R.J.C. (Ed.): Biological Applications of Freezing and Drying. New York: Academic Press 1954.

SPANSWICK, R.M., STOLAREK, J., WILLIAMS, E.J.: The membrane potential of *Nitella translucens*. J. Exp. Botan. **18**, 1–16 (1967).

SPANSWICK, R.M., WILLIAMS, E.J.: Electrical potentials and Na, K, and Cl concentrations in the vacuole and cytoplasm of *Nitella translucens*. J. Exp. Botan. **15**, 193–200 (1964).

STEINHARDT, J., BEYCHOK, S.: Interaction of proteins with hydrogen ions and other smaller ions and molecules, Vol. II, pp. 139–304. In: NEURATH, H. (Ed.): The proteins, New York: Academic Press 1964.

STEVENINCK, R.F.M. VAN: Potassium fluxes in red beet tissue during its "lag phase". Physiol. Plant. **15**, 211–215 (1962).

STEVENINCK, R.F.M., VAN, JACKMAN, M.E.: Respiratory activity and morphology of mitochondria isolated from whole and sliced storage tissue. Australian. J. Biol. Sci. **20**, 749–760 (1967).

STEWART, C.R.: Proline content and metabolism during rehydration of wilted excised leaves in the dark. Plant Physiol. **50**, 679–681 (1972).

STOCKER, O.: Die Dürreresistenz, Vol. III, pp. 696–741. In: RUHLAND, W. (Ed.): Encyclopedia of plant physiology. Berlin-Göttingen-Heidelberg-New York: Springer 1956.

STROGONOV, B.P.: Physiological basis of salt tolerance of plants. Translated from Russian original (1962) by POLJAKOFF-MAYBER, A. and MAYER, A.M. Jerusalem: Israel Program for Scientific Translations 1964.

SUELTER, C.H.: Enzymes activated by monovalent cations. Science **168**, 789–795 (1970).

TAKAOKI, T.: Relationships between plant hydrature and respiration II. J. Sci. Hiroshima Univ. Ser. B: 2, **8**, 73–80 (1957).

TWERSKY, M., FELHENDLER, R.: Effect of water quality on relationships between cationic species and leaf lipids at two development stages in cotton. Physiol. Plant. **29**, 396–401 (1973).

UDOVENKO, G.V., MASHANSKII, V.F., SINITSKAYA, I.A.: Changes of root cell ultrastructure under salinization in plants of different salt resistance. Fiziol. Rast. (Transl.) **17**, 813–818 (1970).

VAADIA, Y., RANEY, F.C., HAGAN, R.M.: Plant water deficits and physiological processes. Ann. Rev. Plant Physiol. **12**, 265–292 (1961).

WAISEL, Y.: Biology of halophytes. New York, London: Academic Press 1972.

WAISEL, Y., ESHEL, A.: Localization of ions in the mesophyll cells of the succulent halophyte *Suaeda monoica* Forssk. by X-ray microanalysis. Experientia **27**, 230–232 (1971).

WALTER, H., KREEB, K.: Die Hydratation und Hydratur des Protoplasmas und ihre ökophysiologische Bedeutung. Protoplasmatologia **II C6**. Vienna: Springer-Verlag 1970.

WEBB, K.L., BURLEY, J.W.A.: Dark fixation of $^{14}CO_2$ by obligate and facultative salt marsh halophytes. Can. J. Botan. **43**, 281–285 (1965).

WEIMBERG, R.: Effect of sodium chloride on the activity of a soluble malate dehydrogenase from pea seeds. J. Biol. Chem. **242**, 3000–3006 (1967).

WEIMBERG, R.: Enzyme levels in pea seedlings grown on highly salinized media. Plant Physiol. **46**, 466–470 (1970).

WINTER, K.: Zum Problem der Ausbildung des Crassulaceensäurestoffwechsels bei *Mesembryanthemum crystallinum* unter NaCl-Einfluß. Planta **109**, 135–145 (1973).

CHAPTER IX

Water Balance and Gas Exchange of Plants under Saline Conditions

J. GALE

A. Water Balance and Salinity

I. The Physiological Drought Hypothesis

Towards the end of the 19th century the reduced growth of plants under conditions of salinity was ascribed to "physiological drought". By this was meant a shortage of water within the plant even when growing under moist but saline soil conditions, or in saline culture solutions (see SCHIMPER, 1898; quoted by STROGONOV, 1964): The lowered osmotic potential of the soil water, resulting from high concentrations of soluble salts, was thought to prevent uptake of water by the plant. Upset water balance was therefore considered to be the main factor in salinity damage, although specific toxic effects were also recognised (cf. BERNSTEIN and HAYWARD, 1958). A corollary to the physiological drought hypothesis was that equi-osmolar concentrations of different salts would have the same relative effect on plant growth.

This hypothesis prevailed for some sixty years (e.g. MAGISTAD, 1945; HAYWARD, 1955) despite the early work of OSTERHAUT (1906) who showed that diluted sea water (a mixed salt solution) was less damaging to plant growth than equi-osmolar concentrations of single salts. Similar conclusions, as to the lower toxicity of mixed salts, were later arrived at by LAGERWERFF and EAGLE (1961).

It has often been observed that the osmotic potential (π) of the sap, expressed from leaves of plants growing under saline conditions, changes in the direction which maintains a constant water potential gradient between leaf and soil (EATON, 1927, 1942; BLACK, 1960; SLATYER, 1961). However, this was not always considered necessarily to disprove the physiological drought hypothesis, as no such adjustment was found in roots. Later, after taking into account "free space", BERNSTEIN (1961, 1963) was able to demonstrate π adjustment in the roots also. It was therefore concluded that water imbalance could not be involved in the salinity response but that the damage (stunted growth, reduced yields, "burnt" or chlorotic leaves etc.) is due rather to the nature of the π adjustment, a concept postulated earlier by VAN DEN BERG (1952). This fitted in well with the common observation that plants grown in saline substrate are often more succulent (e.g. BLACK, 1958) and no less turgid (e.g. GALE et al., 1967) than control plants.

Adjustment of π can be achieved by absorption of a salt which may or may not be toxic (SLATYER, 1961), by the release of K^+ ion from binding sites within the cell, or by hydrolysis of polysaccharides to smaller sugar molecules (BERNSTEIN, 1963).

II. Osmotic Adjustment and the Plant Water Balance

The foregoing emphasis on osmotic adjustment tends to be an oversimplification. The determination of plant water balance, or status, is much more complicated than the maintenance of the osmotic gradient between plant and soil, or between plant and culture solution (COWAN and MILTHORPE, 1968). In the final analysis, water balance is determined by the ratio of uptake to loss and by the immediate history of the plant preceding a particular situation.

Uptake of water is determined by the gradient of total water potential (Ψ_w) between root and soil, and by the resistances to liquid flow in the soil-root system. Loss of water from the plant is determined by the gradient of water potential between leaf and atmosphere or, using a linear function, the gradient of water vapor density between the substomatal leaf cavities and the atmosphere external to the leaf (Δe). As in the liquid phase pathway, rate of loss is modified by the resistances encountered, which in this pathway are the resistances to water vapor loss from the mesophyll walls (r_m), through the stomata (r_s), and through the boundary layer of air around the leaf (r_a).

Non-equilibrium water balance of particular tissues, cells and organelles, is further modified by internal resistances to liquid water flow, and by local generation of changes in Ψ_w-mainly due to changes in π and T (turgor pressure). Furthermore it should be recognised that Ψ_w is composed of three major components — π, T, and τ (matric potential). It is possible to obtain the same equilibrium value of Ψ_w for very different combinations of π, T, and τ, each of which has its own physiological significance. Various combinations of π, T, and τ for the same equilibrium values of Ψ_w are found in different tissues, cells and cellular fractions (NOY MEIR and GINSBURG, 1969).

III. Plant Resistances to Water Movement as Affected by Salinity

One of the primary effects of salt is on the water potential Ψ_w of the soil solution and on the hydraulic conductivity of the soil, both of which affect the availability of water at the soil root interface (SHAINBERG, Chapter III).

Whether or not the plant roots develop a low osmotic potential may be irrelevant to the maintenance of $\Delta \Psi_w$ between soil and root. This is because of the non-osmotic components of Ψ_w in the plant, especially negative values of T in the xylem vessels of the roots, which may result from transpirational pull. In an extreme case, values of P equal to -30 bars were found in the xylem vessels of mangroves, which take up water from the sea by reverse osmosis, despite the very low values of π of their xylem sap (SCHOLANDER et al., 1962).

Salt may also affect hydraulic permeability (L_p) of the roots. Theoretical aspects of this have been discussed by OERTLI and RICHARDSON (1968). O'LEARY (1969), using a pressure bomb technique, measured a considerable decrease in passage of water through roots of red kidney bean grown under saline conditions. KAPLAN and GALE (1972) using the same technique obtained similar results for the halophytic saltbush. HOFFMAN and PHENE (1971) calculated overall plant resistance to liquid water flow, in bean and cotton, as the ratio of (Ψ_w soil $-$ Ψ_w plant) to transpiration. They too found an increased resistance in response to salinity. Similar conclusions were arrived at by KIRKHAM et al. (1969) using a split root technique, with bean plants.

IV. Plant Resistances to Water Vapor Loss as Affected by Salinity

There are many reports of a depression of transpiration under saline conditions [e.g. MEYER (1931) working with cotton, GALE et al. (1967) with onion, beans and cotton, ASHBY and BEADLE (1957), GALE and POLJAKOFF-MAYBER (1970), KAPLAN and GALE (1972) with saltbush, and MEIRI and POLJAKOFF-MAYBER (1970) working with beans]. STROGONOV (1962) differentiates between the effect of sulphate and chloride type salinity on transpiration, finding that the depression of transpiration is greater with the latter.

For some time it was considered possible that the reduction of transpiration by salinity could be caused by the depression of the water vapor density within the subcellular air spaces. This was assumed to result from the presence of dissolved salt in the water phase within the mesophyll walls (e.g. WAISEL, 1972, p. 90). As noted above this would reduce e_{leaf} and hence Δe and transpiration. However, this cannot be more than a negligibly small factor. Owing to the logarithmic relationship between osmotic potential and water activity, even at the highest levels of salinity, e_{leaf} will be depressed by only a few percent[1]. For example when $\pi = -20$ bars, depression of the vapor pressure will be only 1.5% (MILTHORPE, 1962). Furthermore SLATYER (1966) has calculated that, due to back diffusion, salts will be unlikely to concentrate to any great extent within the cell walls at the end of the liquid phase pathway. BERNSTEIN (1971) measured a 3–5 fold increase in solute concentration within the mesophyll cell wall as compared to the concentration in the xylem sap. Even so this concentration was still only 1–7% of that in the cell sap. The accumulation of salt in the mesophyll walls appears therefore to be incapable of bringing about a more than negligible decrease in vapor pressure. However an increase of mesophyll resistance to water vapor loss (r_m) in response to salinity has been reported by KAPLAN and GALE (1972). This increase of r_m was found in *Atriplex halimus* (halophytic saltbush) and attained physiologically significant levels.

The data of KAPLAN and GALE (Table 1) show that salinization produced little or no increase of r_s of the upper leaf surface but a large increase in r_s of the lower leaf surface; the most significant finding was a very large increase in r_m.

The methodology of the measurement of r_m is very problematic (GALE et al., 1967). However in this case there appears to be good evidence for an increase in resistance to water vapor loss in the mesophyll without an equally large effect on resistance to CO_2 fixation. This constitutes incomplete but supporting evidence showing that the mechanism of increased r_m, in response to salt, is not caused by an accumulation of salt at the mesophyll wall interface. If salt was accumulating in the mesophyll cell wall to an extent sufficient to reduce e_l significantly it would be expected that CO_2 uptake (photosynthesis) would be reduced by an even larger

[1] $\pi \propto ln\ e/e_0$ where e_0 is the vapor pressure in salt free water at the same temperature.

Table 1. Photosynthesis, transpiration, and diffusion resistances of leaves of control and salinized *Atriplex halimus* plants. [Adapted from KAPLAN and GALE (1972)]

Parameter	Leaf surface	Control (Knop) culture solution	Salinized culture solution[a]
Photosynthesis (mg dm^{-2} hr^{-1})	Upper	11.7 ± 0.6	10.4 ± 0.4
	Lower	9.7 ± 0.6	6.7 ± 0.6
Transpiration (mg dm^{-2} hr^{-1})	Upper	433 ± 29	289 ± 15
	Lower	750 ± 28	289 ± 15
Stomatal resistance, r_s (sec cm^{-1})	Upper	9.6 ± 1.6	10.7 ± 1.1
	Lower	2.7 ± 0.8	12.5 ± 2.1
Mesophyll resistance, r_m (sec cm^{-1})	Upper and lower	3.4 ± 0.5	13.9 ± 1.9

[a] Osmotic potential of culture solution −10 bars in addition to Knop solution.
Resistances are calculated for water vapor.
Values given are the average of at least six leaves ± standard error.

factor. This is because the function for the reduction of CO_2 solubility in water (a necessary precursor of CO_2 uptake into the wet mesophyll wall) is steeper per unit increase of π than the function for the reduction of e_l (MARKHAM and KOBE, 1941; RATCLIFF and HOLDCRAFT, 1963).

The mechanism for the effect of salt on r_m thus remains at present unclear. If accumulation of salt in the mesophyll wall and consequent reduction of e_l is rejected as a hypothesis, one must conclude that salinity reduces the hydraulic conductivity of the mesophyll cell walls.

Reduction of transpiration under saline conditions, and the increase of r_s, were originally ascribed to reduced turgor and subsequent stomatal closure, resulting from reduction in water uptake. However, increased r_s, in response to salinity, has often been measured concomittantly with an increase in turgor pressure (e.g. GALE et al., 1967). Consequently other explanations, and especially the involvement of plant hormones, were suggested. This is discussed in the following section.

V. Salinity and Plant Hormones

One of the most common effects of salinity is stunting of growth, often without any other sign of damage such as chlorosis or leaf burn. This and other modifications of growth habit, such as increased leafiness (e.g. GALE and POLJAKOFF-MAYBER, 1970), has suggested that growth hormones may be involved in the response of plants to salinity. One of the first hormones to have been studied from this point of view was gibberellic acid. However no relation between gibberellic acid and salinity could be found (NIEMAN and BERNSTEIN, 1959).

Cytokinins are produced in roots and move to the plant tops where they are known to have a function in maintenance, especially maintenance of protein integrity. KAHANE and POLJAKOFF-MAYBER (1968) reported a decrease in amino acid incorporation into protein of pea roots exposed to NaCl or Na_2SO_4. The effect of

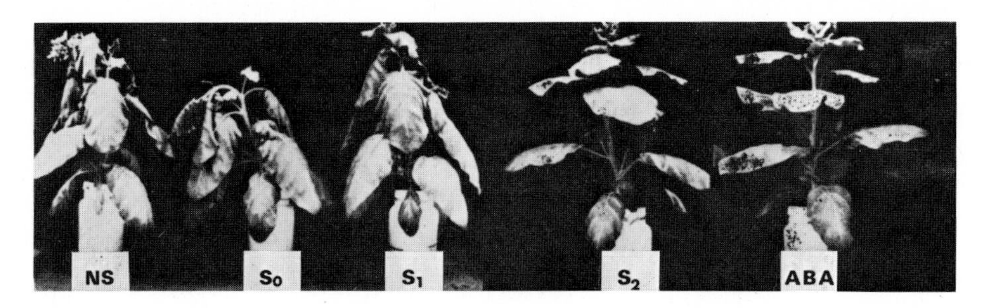

Fig. 1. Appearance of tobacco plants 4 hrs after withholding root aeration. *NS* Non-salinated plants. S_0 NaCl was added when aeration was stopped. S_1 Salinated for 24 hrs before aeration was ceased. S_2 Salinated for 4 days before stopping aeration. *ABA* Abscissic acid, was added to nutrient solution 2 days before stopping aeration. [From MIZRAHI et al. (1972) by courtesy of the authors]

NaCl (but not of Na_2SO_4) could be partially counteracted by exogenously applied kinetin. Kinetin has also been reported to affect the pattern of distribution of Na^+ and Cl^- in bean and cotton plants (EL-SAIDI and KUIPER, 1972). ITAI et al. (1968) showed that in sunflower, bean and tobacco, the level of endogenous cytokinins was reduced under conditions of stress, including water shortage and salinity. This suggests that one reason for metabolic disturbances at the cellular level, under saline conditions, may be lack of cytokinins. However, which is the primary cause and which the effect is not entirely clear.

Plant hormones, especially kinetin and abscissic acid (ABA) appear to play an important role in plant water relations through their effect on stomata. Kinetin promotes, and ABA reduces opening of stomata (LIVNE and VAADIA, 1965, 1972). Recently, a clear relationship has been found between kinetin, ABA and various plant stresses including salinity. MIZRAHI et al. (1972) working with tobacco, found a lowering of the level of endogenous kinetin and an increase of ABA, when the plants were exposed to either mannitol or NaCl (added to the culture solutions). An example from their work, showing the interaction of salinity, lack of aeration and ABA is shown in Fig. 1.

As can be seen in Fig. 1, tobacco plants grown in culture solutions tend to wilt when aeration of the roots is suspended. However, plants grown in poorly aerated saline solutions do not wilt. MIZRAHI et al. found that this effect of salinity was correlated with both an increase in r_s and an increase in the level of endogenous ABA; ABA treated plants grown in normal culture solution, whose aeration was stopped, did not wilt. It should be noted that these experiments were carried out under hot dry conditions. The significance of the environment-salinity interaction as affecting plant growth is discussed in the following chapter.

Other hormones may also be involved in the response of the plant to salinity. For example VERMA et al. (1973) have recently reported that ethylene reversed the inhibition of germination of lettuce seeds caused by salinity. Indole acetic acid (IAA) is probably not directly involved as it does not affect stomatal aperture (JOHANSEN, 1954) although it may affect plant water balance by modifying the number and size of leaves.

VI. The Overall Effect of Salinity on Water Balance and Water Status of Plants

As discussed above, no single parameter such as Ψ_w, π, T, τ or "water saturation deficit" is completely adequate for describing the overall plant water balance and/or status. Plant water potential, Ψ_w, has often been considered to be the most representative parameter of plant water status. However it appears to be of value mainly within the context of liquid water movement where resistances to flow are also defined. It is unlikely that lowered Ψ_w, *per se*, has much influence on metabolic processes such as photosynthesis, as water activity (which is the property to be considered where water enters into biochemical reactions) is only logarithmically related to Ψ_w. Consequently relatively large changes in Ψ_w produce only negligible shifts in water activity (HSIAO, 1973). The same argument holds here as for vapor pressure, with which it is a colligative property (see Section IV, above).

There is some justification for placing special emphasis on π as a measure of water status in relation to slinity. This was what WALTER actually measured for many years in relation to his concept of "Hydrature" (WALTER, 1955). The reason for the special significance of π is that it appears to be highly correlated to physiological functions and to overall growth response. However, by itself, it gives no indication of water balance. It should also be noted that the physiological significance of π depends on the nature of the participating solutes. For example GREENWAY et al. (1972) found that π at -20.4 bars had no effect on respiration of Chlorella, when this level of π was due to the presence of the rapidly permeating, but innocuous, ethylene glycol. The same level of mannitol, which does not readily penetrate the Chlorella cells and which therefore produced dehydration, severely inhibited respiration (Fig. 2). Glucose uptake was inhibited 90% by the mannitol treatment but only 30% by ethylene glycol.

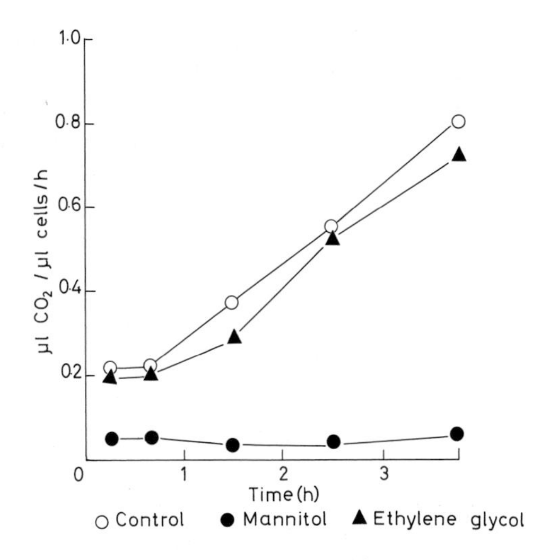

Fig. 2. Effects of various osmotica at −20.4 atm on $^{14}CO_2$ evolution from ^{14}C-glucose (5 mM) by Chlorella pyrenoidosa. [From GREENWAY et al. (1972) by courtesy of the authors]

Table 2. Turgor of control and salinized (NaCl) onion, bean, and cotton plants in a cold and humid or hot and dry atmosphere. [Adapted from GALE et al. (1967)]

Plant	Atmosphere around plants	Salinity level for salinized plants −bars	Turgor pressure −bars	
			Control	Salinized
Onion	Cold humid	−5.4	8.8	2.2
	Hot dry	−5.4	4.8	−3.1
Bean	Cold humid	−3.4	5.7	6.3
Cotton	Hot dry	−5.4	8.0	9.0
Cotton	Cold humid	−8.5	9.7	10.8
	Hot dry	−8.5	6.0	8.0

Turgor pressure T, may be the one parameter which gives an indication of the plant water *balance*. However there are at present no simple, direct methods for measuring T. Estimates of T are usually based on the difference between measurements of Ψ_w and π, both of which involve considerable error, especially measurements of π in salt accumulating plants (KAPLAN and GALE, 1974). This procedure also neglects the contribution of τ.

In view of the above reservations as to methodology and interpretation of measurements of water status, reports on the effect of salt on water balance, which are based on measurements of single parameters should be interpreted with caution. Table 2 brings data (GALE et al., 1967) on turgor of plants grown in the presence or absence of salts, calculated as the difference between measured values of Ψ_w and π.

In most plant species, such as the bean and cotton shown here, osmotic adjustment and reduced transpiration combine to overcome any salt induced increase in root resistance (see also SLATYER, 1961). This often results in a turgor pressure which is equal to or larger, in plants exposed to salinity, than in the control plants. However in some plants (see e.g. data on onion in Table 2) π adjustment is incomplete, and transpiration is not reduced sufficiently to maintain turgor.

The above conclusions as to the effect of salinity on the overall water balance of the plant, are modified when changes occur in the top: root ratio. This ratio is usually lower in plants exposed to salinity (see e.g. BOYER, 1965, for cotton). Plant water balance under conditions of salinity is also affected by other interactive environmental conditions such as weather (see following chapter) or by the salinity regime, i.e. the variation of salt concentration with time.

There are only a few reports on the water balance of plants under a non-steady state salt regime. MEIRI and POLJAKOFF-MAYBER (1969, 1970) have shown, for beans, that in a non-stable salt regime, onset of salinity induced a temporary lag in π adjustment and consequent reduction in the level of Ψ_w and T. They also reported that when salinity was abruptly removed transpiration did not increase to the level of the controls, despite the expected upward surge in turgor pressure, which might have, but evidently did not induce full opening of the stomata.

B. Photosynthesis, Respiration, and Salinity

I. General Considerations and Effect of Salinity on Photosynthesis

Evaluation of the effect of salinity on net photosynthesis (P) and on dark respiration (R_D) is much complicated by the variety of methods and bases of calculation used by different workers. Photosynthesis, net or gross, has been expressed per unit — leaf area (e.g. GALE and POLJAKOFF-MAYBER, 1970), leaf fresh weight (e.g. LAPINA and POPOV, 1970) or per plant. Attention must be given to the experimental conditions during measurement of gas exchange as they may be very varied. For example NIEMAN (1962) measured the effect of sodium chloride salinity on photosynthesis of twelve plant species in a Warburg apparatus with illumination of 1000 f.c. This procedure would reveal any effect of salinity on the light dependent reactions of photosynthesis. It would not however be sensitive to small changes induced by salt in the "dark" CO_2 fixation pathway of photosynthesis, as under the above conditions light would usually be a limiting factor. The period of exposure to salt, and age of leaf etc. also varies greatly in different reports and frequently makes comparison of results difficult.

Most workers have measured P by the rate of net CO_2 uptake. Recent work by CHIMIKLIS and KARLANDER (173) with the alga *Chlorella*, has shown that P as measured by O_2 evolution may be increased while net CO_2 fixation is decreased in response to salinity.

Bearing in mind the above reservations, the commonly found effect of salinity of whatever variety of salt, is a decrease in P per unit leaf area. This has been reported by a number of workers. For example, GALE et al. (1967) reported such an effect, for NaCl, on onions, beans and cotton; so also: LAPINA and POPOV (1970) working with NaCl and tomato; GALE and POLJAKOFF-MAYBER (1970) working with NaCl, Na_2SO_4, and saltbush; UDOVENKO et al. (1971) working with NaCl, Na_2SO_4, and barley, wheat and bean; HOFFMAN and PHENE (1971) working with NaCl and bean and cotton; and LAPINA and BIKMUKHAMETOVA (1972) working with Na_2SO_4, NaCl and corn. An exception to the general rule can be seen in halophytes, where low concentrations of salt do not always reduce, and may even enhance, photosynthesis (GALE and POLJAKOFF-MAYBER, 1970). In general, photosynthesis is reduced in proportion to salt concentration. The percentage reduction varies greatly in the different plant species and varieties.

There is no general concensus as to the relative reduction of photosynthesis caused by the two main types of salinity — NaCl and Na_2SO_4. For example, UDOVENKO et al. (1971) found no significant difference between the effect of the two salts on barley, wheat or beans, while LAPINA and BIKMUKHAMETOVA (1972) found no effect of low concentrations (up to -4.4 bars π) of Na_2SO_4 on P of corn, whereas NaCl decreased P at all concentrations. The reverse was reported by GALE and POLJAKOFF-MAYBER (1970) working with *Atriplex halimus*. They found that the effect of NaCl on P was negligible with concentrations up to -10 bars and there was only a small effect with concentrations up to -20 bars. In contrast, Na_2SO_4 reduced P relative to the control even at -1 bar. It should be noted that the above reports refer to different plant species. Neither were the experimental conditions identical.

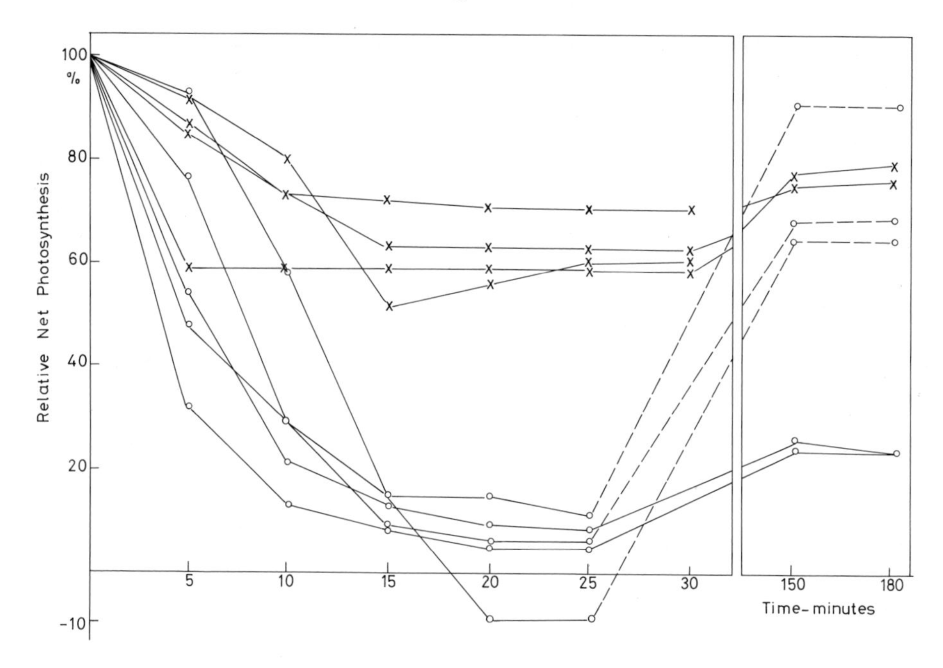

Fig. 3. Effects of elevated air temperature and desalination of the nutrient solution on net photosynthesis of salinated onion plants. Abscissa: time in minutes after elevation of air temperature from 25° C to 35° C. Ordinate: net photosynthesis at 35° C in per cent of that of the same plant at 25° C. × – × controls; o – o salinated plants; o – – o salinated plants after the culture solutions had been changed to normal $^1/_2$ Hoagland (−0.4 atm). Dew point constant at −22° C. [From GALE et al. (1967)]

In conclusion, although there is little doubt that photosynthesis is reduced as salinity increases the degree and mechanism of this effect may vary greatly in different plant species even with the same type of salt.

The considerable difference in species response is illustrated in the work of GALE et al. (1967) for three plant species: onion, bean and cotton. In onion plants the main mechanism for reduction of *P* (per unit area) appeared to be closure of stomata due to water imbalance (Fig. 3). The rate of *P* of the onion plants, grown with NaCl in the culture solution, was greatly reduced (in comparison to the controls) when the plants were exposed to hot dry conditions. This tendency was reversed when the plants were allowed to absorb water by transfer to non-saline solutions. In other words, although the plants remained essentially "salinated", the boosted turgor pressure evidently caused the stomata to reopen, allowing the basically unaffected *P* to rise to the level of the controls. Such an effect was not found in either the bean or the cotton plants. Furthermore, measurement of the response of bean and cotton plants to varying levels of carbon-dioxide concentration and light intensity, showed that the reduction of *P* found in plants grown under saline conditions, was due in each species to a different mechanism. In the bean plants, within the first few days after being placed in saline culture solutions, reduction of *P* appeared to be due mainly to interference with CO_2 uptake. This was caused by incomplete opening of the stomata despite the high leaf turgor. High concentrations of CO_2 could overcome this effect on P. In the cotton plants, reduction of P was greater at hight than at low concentrations of CO_2, indicating damage to the light reactions or to the biochemical reactions of CO_2 fixation.

Very little is known on the cellular mechanisms involved in the reduction of P caused by salinity. Changes in fine-structure of chloroplasts in response to NaCl salinity have been reported by BLUMENTHAL-GOLDSCHMIDT and POLJAKOFF-MAYBER (1968) in *Atriplex halimus* and by LAPINA and POPOV (1970) in tomatoes. The changes observed in *Atriplex* occurred at about the same level of π, at which reduction of P was found by GALE and POLJAKOFF-MAYBER (1970).

SANTARIUS and RENATE (1967) studied the effect of dehydration, caused by either sugars or NaCl, on chloroplasts isolated from spinach or beet. Dehydration by sugars, to 15% of initial water content, had little effect on Hill reaction or photophosphorylation. However dehydration by NaCl caused a reduction in the two processes. The reduction was reversible at low and irreversible at high concentrations of salt. From the work of SANTARIUS and RENATE (1967), PLAUT (1971) and others, it appears that although the photosynthetic system is resistant to a considerable degree of dehydration, it may become damaged if the dehydration is accompanied by an increase in electrolyte concentration.

Salinity may also affect photosynthesis by modifying the optical properties of leaves. For example CARTER and MYERS (1963) reported that reflectance of leaves of grapefruit trees, growing in orchards having saline soils, was as much as doubled at wavelengths around 650 nm. UDOVENKO et al. (1971) found that in plants grown under saline conditions, chloroplasts sank to lower positions within the pallisade cells. The significance of these last two observations is not clear and will probably vary, depending on other environmental conditions such as light intensity, but they do illustrate once again the many-faceted nature of the salinity response.

II Salinity and Respiration

The above discussion (Section B. I.) refers mainly to Net-Photosynthesis. However dark respiration (R_d) may also take place during photosynthesis. Whether or not R_d continues in the light at the same rate as in the dark is still open to debate. However there is no doubt that increased R_d during the dark will decrease net-P calculated on the basis of the full 24-hrs cycle. Salinity often affects the rate of respiration, and so must be taken into account in any analysis of the effects of salinity on carbon fixation.

Plants having the C_3-carboxylation pathway of CO_2 fixation also exhibit photorespiration (R_p), which releases CO_2 (ZELITCH, 1971). R_p may also be present in plants having the C_4-carboxylation pathway where it is more difficult to detect. As R_p much reduces net vs gross photosynthesis any influence salinity may have on R_p would also affect net-P. However in contrast to R_d there appears to have been no report to date on the effect of salinity on R_p. Even so this possibility should be investigated as R_p is responsible for the release of a very considerable fraction of the CO_2 fixed in photosynthesis.

One of the dominant theories explaining the general stunting of growth, in response to salinity, relates growth reduction to the diversion of available energy from accumulative growth (entropy reduction) to maintenance. Energy requirements are assumed to be increased due to the pumping of ions agains electrochemical-potential gradients and also to the rebuilding of organelles and protein units whose rate of disruption may be increased by the presence of high concentrations of electrolytes.

The phenomenon of "salt respiration"-increased respiration of roots in the presence of high concentrations of mineral salts-has been recognised for many years (see RAINS, 1972). High concentrations of NaCl have often, but not always (cf. NIEMAN, 1962) been reported to increase respiration of roots and other tissues. For example, LIVNE and LEVIN (1967) working with pea seedlings reported that R_d in leaves of pea seedlings grown in saline solution (77 mM NaCl) was stimulated by about 10–15% in the tops and by 33% in the roots.

As with most other salinity responses, there appear to be large differences between species. For example LAPINA and POPOV (1970) found reduced R_d in tomato plants grown in culture solutions containing NaCl, whereas LAPINA and BIKMUKHAMETOVA (1972) found increased R_d in corn grown in the presence of either Na_2SO_4 or NaCl.

LESSANI and ANDREOPOULOS-RENAUD (1969) showed that whether R_d is stimulated or repressed by salinity may be a question of salt concentration. They grew lucerne in culture solutions to which NaCl was added at concentrations of from 0 to 12 g/l. Leaf respiration was measured in a Warburg apparatus. The same pattern of R_d response was obtained whether the tests were made with neutral or with equi-saline solutions. R_d of tissue samples, taken from plants grown in culture solutions of increasing concentrations between 0 and 5 g/l NaCl, was higher the greater the concentration of salt. R_d of tissue from plants grown at 5 g/l NaCl was ~40% greater than in the controls. Higher levels of salinity decreased respiration. Respiration of plants grown with 12 g/l NaCl was 10% less than in the controls. Plants grown in normal culture solution did not respond to salt added in the Warburg apparatus. The data of HOFFMAN and PHENE (1971) for kidney bean also show a larger stimulation of R_d at low (−2.4 bars) than at high (−4.4 bars) levels of salinity: ~80% and 10% respectively. However in their work with cotton under conditions of −6.4 bars salinity, R_d was increased by ~60% and under −12.5 bars salinity by ~85% *vs.* the controls.

The increase of R_d in response to salt is probably a truly adaptive and not a pathological reaction (which would merely reduce overall net CO_2 fixation). Evidence for this, apart from the above teleological argument, is seen in the work of MOROZOVSKII and KABANOV (1970) and LIVNE and LEVIN (1967). The former authors report that the RQ (CO_2 given off/O_2 taken up) remained constant in NaCl salinized plants. LIVNE and LEVIN working with mitochondria isolated from NaCl salinized pea seedlings, found a 25–75% stimulation of R_d in response to salt, while the phosphorylation/oxidation ratio (P/O) remained constant. In these two works, respiration remained normal, under conditions of salinity, despite change in rate. There was no increase of uncoupling and the plants were evidently utilizing the energy released and were reconverting ATP to ADP. UDOVENKO et al. (1972) found an increase of R_d in response to salt that was

Table 3. Effect of NaCl salinity on the 24 hrs CO_2 balance of bean and cotton plants. [Adapted from HOFFMAN and PHENE (1971)]

Salinity	Leaf area	Net CO_2 fixed during light period	CO_2 evolved during dark period	Net CO_2 assimilation per 24 hrs
–bars	dm^2/plant	mg $dm^{-2} \cdot day^{-1}$	mg $dm^{-2} \cdot day^{-1}$	mg $dm^{-2} \cdot day^{-1}$
Kidney Bean				
0.4	28.6	78.1	17.7	60.4
2.4	11.7	66.1	28.0	38.1
4.4	9.0	48.5	20.7	27.8
Cotton				
0.4	30.4	57.0	10.7	46.3
6.4	23.8	44.1	15.5	28.6
12.4	18.3	41.4	18.9	22.5

greater in a salt resistant than in a salt sensitive species. In both species efficiency of respiration was reduced under saline conditions. Their criterion of "efficiency" was the increase in R_d which could be produced by the uncoupler DNP; a criterion concerning which they themselves have reservations.

The work of HOFFMAN and PHENE (1971), who measured gas exchange of whole bean and cotton plants throughout the day, demonstrate the very large effect that increased R_d can have on net 24-hrs CO_2 fixation. Table 3 shows some of their data.

The data shown in Table 3 demonstrate that increased R_d can constitute a major factor in the reduced overall leaf fixation of CO_2 found under conditions of salinity. Note that the figures for "CO_2 fixed during the light period" are also net, not gross, photosynthesis. Net photosynthesis includes any dark respiration which may be taking place during periods of photosynthesis. This reduced fixation per unit leaf area, compounded by the reduced leaf area, is responsible for much of the overall retarded growth. It should be noted that the reduced leaf area may also be, in part, the result of lowered net photosynthesis, as leaf expansion is itself dependent on net photosynthesis.

III. Changes in the Pathway of CO_2 Fixation in Response to Salinity

There have recently been a number of reports of changes in metabolic pathways in response to salt. Some of these affect gas exchange. The induction of a partial switch in the biochemical pathways of respiration has been discussed in the chapter by KYLIN and QUATRANO (Chapter 8).

It is difficult at present to know whether these metabolic changes are merely passive or pathological responses, or whether they are adaptive. By "adaptive response" is meant a change in behavior which confers an advantage in coping with the saline situation.

A number of the reports of switches in metabolism, which affect gas exchange, refer to plants having the Crassulaceaen type acid-metabolism (CAM). In these plants CO_2 is taken up and fixed temporarily in the dark, in organic acids. During the daytime light period the CO_2 is released within the leaf and refixed and reduced to carbohydrates and to other energy-rich materials, by photosynthesis. This is advantageous in water-scarce situations as it enables the stomata to be closed, and water conserved, during the daylight hours. In a humid but saline habitat, the reduction of the turnover of water (and, of necessity, of the uptake of passively imbibed salts) may also be considered to be beneficial. The disadvantage of this system is its relatively low capacity for dark CO_2 fixation.

KARMARKAR and JOSHI (1969) showed how organic acid synthesis (dark CO_2 fixation) is stimulated in the CAM plant *Bryophyllum pinnatum* by the presence of small quanties (0.04 M) of NaCl. They showed that this was due to the Cl^- ion and not to either Na^+ or $SO_4^=$. Recently, WINTER and WILLERT (1972) and WINTER (1973) have demonstrated activation of the dark fixation pathway, in response to NaCl in the CAM plant *Mesembryanthemum crystallinum.*

Non-Crassulaceaen terrestrial and marine plants also have a small capacity for fixing CO_2 in the dark, which may be affected by presence of salt. JOSHI et al. (1962) showed how marine plants (i.e. plants growing in a highly saline environment) incorporate CO_2 in the dark, mainly into amino acids, whereas spinach leaves growing on non-saline substrates incorporate CO_2 into organic acids. In the presence of salts (0.22 M NaCl in the homogenate) spinach also incorporated CO_2 into amino acids.

CAM type metabolism is similar in many of the enzymes involved to the phosphoenolpyruvate (PEP)C_4-carboxylation pathway (ZELITCH, 1971). It is therefore interesting to find that there is also some evidence for a shift from C_3-carboxylases (initially Ribulose di-phosphate-RuDP) to the PEP-C_4 pathway, in response to salinity, especially in halophytic plants. This phenomenon has been reported for *Sueda fructicosa* by AHMAD and HEWITT (1971) and for the highly salt tolerant *Aeluropus litoralis*, by SHOMER-ILAN and WAISEL (1973). The latter authors also claim some evidence for the same phenomenon in *Zea mays* and *Chloris gayana.*

It should however be noted that halophytic plants have a small but definite salt requirement for optimal growth (cf. WAISEL, 1972) and some of the reported responses may be merely a reflection of this requirement. For example, in the above work with *Aeluropus*, PEP carboxylase activity was stimulated by salt by a factor of 8 but, at the same time, RuDP carboxylase activity also increased by a factor of 3.

C. Summary

Before summarizing the effects of the above factors of gas exchange, a short word of caution: gas exchange is usually measured per unit leaf area, and this is often the basis for assessing the effect of salinity. However, salt can also modify the leaf area of the plant, either diminishing or increasing the total area. An example of this interaction can be seen in the report by GREENWAY (1968) on *Atriplex nummurlaria* where low concentrations of salt (100 meq/l NaCl) did not substantially affect the net assimilation rate, but leaf area per plant was much increased. The result was a stimulation of overall growth in response to salinity. The same effect was found by GALE and POLJAKOFF-MAYBER (1970) for *Atriplex halimus*. A similar difference in modification of photosynthesis and transpiration as expressed per unit leaf area or per plant, was shown for two species of *Atriplex*, one with the C_3 and one with the C_4, CO_2 fixation pathways, by SLATYER (1970).

Under saline conditions, the driving force for water uptake ($\Delta\Psi_w$) is usually, if not always, maintained. This is brought about by osmotic adjustment of the plant tissues. However, at the same time, there is an increase in the resistance to water flow both in the soil and in the root system.

The relationship between plant water balance and salinity cannot be resolved by a study of osmotic adjustment alone, as was implicit in the original "physiological drought" concept. Reference must be made to the effect of salinity on each of the many factors governing the entry, passage and evaporation of water, into, through and from the plant. Furthermore, no single parameter, such as osmotic potential (π) or turgor potential (T), should be used alone for evaluating plant water status.

The overall resistance to loss of water vapor from the leaves, r_l, increases under conditions of salinity. This can be related to an increase in stomatal resistance, r_s, for which there is much evidence, and to an increase in mesophyll resistance, r_m, for which evidence has only recently been forthcoming. It should however be recognised that much of the evidence for increased r_s is based on methodology which, a-priori, assumes zero r_m. Overall resistance, r_l, is usually measured without it being partitioned into r_m and r_s.

Changes in the level of hormones also appear to be involved in the response of plants to salinity, especially by way of their effect on plant water balance. The level of endogenous kinetin (which promotes opening of stomata) is lowered while that of abscissic acid (ABA) (which depresses opening of stomata) is raised under conditions of salinity. However, as with all other effects of hormones, little is known of the mechanism of their action other than their general influence on protein integrity (Kinetin) stomatal aperture (Kinetin and ABA) and number and area of the leaves (Indole acetic acid).

In conclusion it appears that under saline conditions the water potential gradient from soil solution to plant is maintained. Furthermore there is an increase in the leaf resistance to water vapor loss; this will tend to counteract the effect of any increase in resistance to water flow occurring in the roots. The result is both a high turgor and a high osmotic concentration in plants grown under saline conditions. This increase of osmotic concentration may in itself be detrimental. The overall rate of water turnover (uptake and transpiration) is generally reduced.

The above appears to hold true with the reservations that: (a) there is not full osmotic adjustment in every species; (b) fluctuating levels of salinity will produce transient changes in plant water balance, and (c) modifications in plant morphology, such as leafiness and top to root ratio, will also affect the overall plant water balance.

Photosynthesis is nearly always reduced by salinity, of whatever type. The concentration at which salinity brings about severe reduction in photosynthesis varies greatly in different plant species. Low levels of salinity may even enhance photosynthesis of halophytes. There are also large differences in response to the various types of salt, such as NaCl and Na_2SO_4. Which step of the photosynthesis process is affected by salinity may also vary considerably.

One certain way in which salinity affects photosynthesis is by reducing stomatal aperture. This interferes with CO_2 diffusion into the plant. The result is a reduction of photosynthesis whenever CO_2 is a limiting factor. CO_2 concentration may be the limiting factor whenever light intensity is high and the basic light reactions and biochemical pathways of photosynthesis have not been damaged. As discussed above, partial closure of stomata is often found in plants exposed to salinity, even when there is a full adjustment of the internal osmotic concentration, and turgor is high.

When the level of salinity is high or the plant species is particularly sensitive, or the exposure to salt has been protracted, breakdown of the photosynthetic apparatus becomes increasingly severe. This is expressed at the whole plant level by leaf chlorosis and, finally, by necrosis and a parallel reduction in photosynthesis per unit leaf area.

Under saline conditions plants require more energy for pumping ions against electro-chemical gradients and for maintenance; this energy appears to be supplied by an increase of respiration but also, possibly, directly from photo-phosphorylation. The increase of respiration and use of energy derived directly from photosynthesis is correlated to a decrease in CO_2 fixation and in overall plant growth. At very high levels of salinity, respiration is reduced, this effect being more pronounced in salt sensitive species. As a result there may be a shortage of energy for maintenance at the very time when demand is greatest.

There appears to be some evidence of a shift from the C_3 pathway of CO_2 fixation to the Crassulaceaen acid type (CAM) metabolism, in response to salinity. Present reports of this phenomenon refer to halophytes. The shift to the CAM pathway may be an actively adaptive response, which confers ecological advantage. This is because turnover of water, and hence passive uptake of salt, is much reduced in CAM type plants.

References

AHMAD, R., HEWITT, E.J.: Studies on the growth and phosphatase activities in *Suaeda fructicosa*. Plant and Soil **34**, 691–696 (1971).

ASHBY, W.C., BEADLE, N.C.W.: Studies in halophytes. III. Salinity factors in the growth of Australian salt-bushes. Ecology **38**, 344–352 (1957).

BERG VAN DEN, C.: The influence of absorbed salts on growth and yield of agricultural crops on salty soils. Versl. Land. Grav. **58**, 5 (1952).

BERNSTEIN, L.: Osmotic adjustment of plants to saline media. I. Steady State. Am. J. Botan. **48**, 908–918 (1961). II. Dynamic Phase. Am. J. Botan. **50**, 360–370 (1963).

BERNSTEIN, L.: Calcium and salt tolerance of plants. Science **167**, 1387 (1970).

BERNSTEIN, L.: Method for determining solutes in the cell walls of leaves. Plant Physiol. **47**, 361–365 (1971).

BERNSTEIN, L., HAYWARD, H.E.: Physiology of salt tolerance. Ann. Rev. Plant Physiol. **9**, 25–46 (1958).

BLACK, R.F.: Effect of sodium chloride on leaf succulence and area of *Atriplex hastata* L. Australian J. Botan. **6**, 306–321 (1958).

BLACK, R.F.: Effects of NaCl on the ion uptake and growth of *Atriplex vesicaria* Heward. Australian J. Biol. Sci. **13**, 249–266 (1960).

BLUMENTHAL-GOLDSCHMIDT, S., POLJAKOFF-MAYBER, A.: Effect of substrate salinity on growth and on sub-microscopic structure of leaf cells of *Atriplex halimus* L. Australian J. Botan. **16**, 469–478 (1968).

BOYER, S. J.: Effects of water stress on metabolic rates of cotton plants with open stomata. Plant Physiol. **40**, 229–34 (1965).

CARTER, D. L., MYERS, V. I.: Light reflectance and chlorophyll and carotene contents of grapefruit leaves as affected by Na_2SO_4, NaCl and $CaCl_2$. Proc. Am. Soc. Hort. Sci. **82**, 217–221 (1963).

CHIMIKLIS, P. E., KARLANDER, E. P.: Light and calcium interactions in *Chlorella* inhibited by sodium chloride. Plant Physiol. **51**, 48–56 (1973).

COWAN, I. R., MILTHORPE, F. L.: Plant factors influencing the water status of plant tissues, Vol. I, 107–136. In: KOZLOWSKI, T. T. (Ed.): Water deficits and plant growth. New York: Academic Press 1968.

EATON, F. M.: The water requirement and cell sap concentration of Australian saltbush and wheat as related to the salinity of the soil. Am. J. Botan. **14**, 967–972 (1927).

EATON, F. M.: Toxicity and accumulation of chloride and sulphate in plants. J. Agr. Res. **66**, 357–399 (1942).

EL-SAIDI, M. T., KUIPER, P. J. C.: Effect of applied kinetin on uptake and transport of ^{22}Na and ^{36}Cl in bean and cotton plants. Meded. Landbouwhogesch. Wageningen **72**, 1–5 (1972).

GALE, J., KOHL, H. C., HAGAN, R. M.: Changes in the water balance and photosynthesis of onion, bean, and cotton plants under saline conditions. Physiol. Plant. **20**, 408–420 (1967).

GALE, J., POLJAKOFF-MAYBER, ALEXANDRA.: Interrelations between growth and photosynthesis of saltbush (*Atriplex halimus* L.) grown in saline media. Australian J. Biol. Sci. **23**, 937–945 (1970).

GREENWAY, H.: Growth stimulation by high chloride concentrations in halophytes. Israel J. Botan. **17**, 169–177 (1968).

GREENWAY, H., LANGE, B., LEAHY, M.: Effects of rapidly and slowly permeating osmotica on macromolecules and sucrose synthesis. J. Exp. Botan. **23**, 459–468 (1972).

HAYWARD, H. E.: Factors affecting the salt tolerance of horticultural crops. Report XIV Internat. Hort. Congr. 385–399 (1955).

HOFFMAN, G. J., PHENE, C. J.: Effect of constant salinity levels on water use efficiency of bean and cotton. Trans. Am. Soc. Agr. Eng. **14**, 1103–1106 (1971).

HSIAO, T. C.: Plant responses to water stress. Ann. Rev. Plant Physiol. **24**, 519–570 (1973).

ITAI, C., RICHMOND, A., VAADIA, Y.: The role of root cytokinins during water and salinity stress. Israel J. Botan. **17**, 187–195 (1968).

JOHANSEN, S.: Effect of indole-acetic acid on stomata and photosynthesis. Physiol. Plant. **7**, 531–537 (1954).

JOSHI, G., DOLAN, T., GEE, R.; SALTMAN, P.: Sodium chloride effect on dark fixation of CO_2 by marine and terrestrial plants. Plant Physiol. **37**, 446–449 (1962).

KAHANE, I., POLJAKOFF-MAYBER, ALEXANDRA.: Effect of substrate salinity on the ability for protein synthesis in pea roots. Plant Physiol. **43**, 1115–1119 (1968).

KAPLAN, A., GALE, J.: Effect of sodium chloride salinity on the water balance of *Atriplex halimus*. Australian J. Biol. Sci. **25**, 895–903 (1972).

KAPLAN, A., GALE, J.: Modification of the pressure-bomb technique for measurement of osmotic potential in halophytes. J. Expt. Botan. **25**, 663–668 (1974).

KARMARKAR, S. M., JOSHI, G. V.: Effect of sand culture and sodium chloride on growth, physical structure and organic acid metabolism in *Bryophyllum pinnatum*. Plant and Soil **30**, 41–48 (1969).

KIRKHAM, M. B., GARDNER, W. R., GERLOFF, G. C.: Leaf water potential of differentially salinized plants. Plant Physiol. **44**, 1378–1382 (1969).

LAGERWERFF, J. V., EAGLE, H. E.: Osmotic and specific effects of excess salts on beans. Plant Physiol. **36**, 472–477 (1961).

LAPINA, L. P., BIKMUKHAMETOVA, S. A.: Effect of isoosmotic concentrations of sodium sulphate and chloride on photosynthesis and respiration in corn leaves. Soviet Plant Physiol. **19**, 792–797 (1972).

LAPINA, L. P., POPOV, B. A.: Effect of sodium chloride on the photosynthetic apparatus of tomatoes. Soviet Plant Physiol. **17**, 477–481 (1970).

LESSANI, M. H., ANDREOPOULOS-RENAUD, U.: Effet de la présence du chlorure de sodium dans le milieu sur l'activité respiratoire et sur le taux des glucides et des acides organiques chez la lucerne. C. R. Acad. Sci. Paris **269**, 951–953 (1969).

LIVNE, A., LEVIN, N.: Tissue respiration and mitochondrial oxidative phosphorylation of NaCl-treated pea seedlings. Plant Physiol. **42**, 407–414 (1967).

LIVNE, A., VAADIA, Y.: Stimulation of transpiration rate in barley leaves by kinetin and gibberellic acid. Physiol. Plant. **18**, 658–664 (1965).

LIVNE, A., VAADIA, Y.: Water deficits and hormone relations, Vol. III, pp. 255–276, In: KOZLOWSKI, T. T. (Ed.): Water deficits and plant growth. New York: Academic Press 1972.

MAGISTAD, O. C.: Plant growth relations on saline and alkaline soils. Botan. Rev. **11**, 181–230 (1945).

MARKHAM, A. E., KOBE, K. A.: The solubility of carbon dioxide and nitrous oxide in aqueous salt solutions. J. Am. Chem. Soc. **63**, 449–454 (1941).

MEIRI, A., POLJAKOFF-MAYBER, ALEXANDRA.: Effect of variations in substrate salinity on the water balance and ionic composition of bean leaves. Israel J. Botan. **18**, 99–112 (1969).

MEIRI, A., POLJAKOFF-MAYBER, A.: Effect of various salinity regimes on growth, leaf expansion and transpiration rate of bean plants. Soil Sci. **109**, 26–34 (1970).

MEYER, B. S.: Effects of mineral salts upon the transpiration and water requirement of the cotton plant. Am. J. Botan. **18**, 79–93 (1931).

MILTHORPE, F. L.: Plant factors involved in transpiration, in UNESCO Arid Zone Research XVI, 107–116, UNESCO (1962).

MIZRAHI, Y., BLUMENFELD, A., RICHMOND, A. E.: The role of abscissic acid and salination in the adaptive response of plants to reduced aeration. Plant and Cell Physiol. **13**, 15–21 (1972).

MIZRAHI, Y., RICHMOND, A. E.: Hormonal modification of plant response to water stress. Australian J. Biol. Sci. **25**, 437–442 (1972).

MOROZOVSKII, V. V., KABANOV, V. V.: Efficiency of respiration in pea and glasswort under NaCl salinization of the substrate. Soviet Plant Physiol. **17**, 482–486 (1970).

NIEMAN, R. H.: Some effects of sodium chloride on growth, photosynthesis and respiration of twelve crop plants. Botan. Gaz. **123**, 279–285 (1962).

NIEMAN, R. H., BERNSTEIN, L.: Interactive effects of gibberellic acid and salinity on the growth of beans. Am. J. Botan. **46**, 667–670 (1959).

NOY-MEIR, I., GINZBURG, B. Z.: An analysis of the water potential isotherm in plant tissue II. Comparative studies on leaves of different types. Australian J. Biol. Sci. **22**, 35–52 (1969).

OERTLI, J. J., RICHARDSON, W. F.: Effects of external salt concentrations on water relations in plants. IV. The compensation of osmotic and hydrostatic water potential differences between root xylem and external medium. Soil Sci. **105**, 177–183 (1968).

O'LEARY, J. W.: The effect of salinity on permeability of roots to water. Israel J. Botan. **18**, 1–9 (1969).

OSTERHOUT, W. J. V.: On the importance of physiologically balanced solutions for plants. Botan. Gaz. **42**, 127–134 (1906).

PALLAGHY, C. K.: Salt relations of *Atriplex* leaves, pp. 57–62. In: JONES, R. (Ed.): The biology of *Atriplex*. CSIRO Australia (1970).

PLAUT, Z.: Inhibition of photosynthetic carbon-dioxide fixation in isolated spinach chloroplasts exposed to reduced osmotic potential. Plant Physiol. **48**, 591–595 (1971).

RAINS, D. W.: Salt transport by plants in relation to salinity. Ann. Rev. Plant Physiol. **23**, 367–388 (1972).

RATCLIFF, G. A., HOLDCRAFT, J. G.: Diffusivities of gases in aqueous electrolyte solutions. Trans. Inst. Chem. Engnrs. **41**, 315–319 (1963).

SANTARIUS, K. A., RENATE, E.: Das Verhalten von Hill-reaktion und Photophosphorylierung isolierter Chloroplasten in Abhängigkeit vom Wassergehalt. I. Wasserentzug mittels konzentrierter Lösungen. Planta **73**, 91–108 (1967).

SCHOLANDER, P. F., HAMMEL, H. T., HEMMINGSEN, E., GAREY, W.: Salt balance in mangroves. Plant Physiol. **37**, 722–729 (1962).

SHOMER-ILAN, A., WAISEL, Y.: The effect of sodium chloride on the balance between the C_3– and C_4-carbon fixation pathways. Physiol. Plant. **29**, 190–193 (1973).
SLATYER, R. O.: Effect of several osmotic substrates on the water relations of tomato. Australian J. Biol. Sci. **14**, 519–540 (1961).
SLATYER, R. O.. Some physical aspects of non-stomatal control of leaf transpiration. Agr. Met. **3**, 281–292 (1966).
SLATYER, R. O.: Comparative photosynthesis growth and transpiration of two species of *Atriplex*. Planta **93**, 175–189 (1970).
STROGONOV, B. P.: Physiological basis of salt tolerance of plants (as affected by various types of salinity) (1962). English Edition Jerusalem: IPST (1964).
UDOVENKO, G. V., KHAROVA, G. V., NASTINOVA, G. E.: Changes in the rate and energy efficiency of respiration in plants of different salt resistance under conditions of salinity. Soviet Plant Physiol. **19**, 676–82 (1972).
UDOVENKO, G. V., SEMUSHINA, L. A., PETROCHENKO, N. G.: Character and probable causes of the changes in the photosynthetic activity of plants during salinization. Soviet Plant Physiol. **18**, 598–604 (1971).
VERMA, C. M., BOHRA, S. P., SANKHLA, N.: Lettuce seed germination: Reversal of salinity induced inhibition by ethylene. Current Sci. **42**, 294–295 (1973).
WAISEL, Y.: Biology of halophytes. New York: Academic Press 1972.
WALTER, H.: Water economy and the hydrature of plants. Ann. Rev. Plant Physiol. **6**, 239–252 (1955).
WINTER, K.: CO_2-Fixierungsreaktionen bei der Salzpflanze *Mesembryanthemum crystallinum* unter variierten Außenbedingungen. Planta **114**, 75–85 (1973).
WINTER, K., WILLERT, D. J.. Sodium chloride induced crassulaceaen acid metabolism in *Mesembryanthemum crystallinum*. Z. Pflanzenphysiol. **67**, 166–170 (1972).
ZELITCH, I.: Photosynthesis, photorespiration and plant productivity. New York: Academic Press 1971.

CHAPTER X

The Combined Effect of Environmental Factors and Salinity on Plant Growth

J. GALE

A. General Considerations

In this chapter the combined effect of the aerial environment and salinity on plant growth is examined. Aspects of soil environment salinity interactions are discussed by PECK (Chapter V).

For many years it has been noted that the relative effect of salinity on plant growth is often, if not always, a function of weather (TRELEASE and LIVINGSTON, 1924; AHI and POWERS, 1938; WALL and HARTMAN, 1942; BERNSTEIN and AYERS, 1951). Generally an increase of the "severity" of the weather (heat, wind, dryness) is associated with increased severity of the symptoms of salt damage.

A considerable difficulty in any analysis of weather effects is the non-specific nature of many of the component factors, such as temperature, and their mutual interactions. For example, radiation is the driving force of photosynthesis, but also of transpiration where it operates in conjunction with many other factors, such as air temperature, humidity, wind and turbulence. A second difficulty is the failure of many reports to differentiate between a genuine interaction and a mere summation of the effects of salt and other factors on plant growth. This has been pointed out by LUNT et al. (1960a, b) who analyze their own data, on the interaction of environment and salt, on the basis of Baule's principle. Baule's principle states that the resultant of two factors is their product. For example, if extreme heat reduces growth by 20% and excess salt by 30%, then together, without any special interaction, they will reduce growth to $80\% \times 70\% = 56\%$ of the controls (i.e. a reduction of 44%). In this case a reduction of growth equal to or less than 44% would not indicate any interaction of heat and salinity.

The above two reservations should be borne in mind when considering many of the reports discussed below.

B. Temperature and Salinity

Temperature is the least specific of all the environmental factors as it affects movement of salts in the soil, uptake of salts, overall biochemical processes in the plant, and transpiration.

AHI and POWERS (1938) grew plants in diluted sea water and obtained much greater growth in a cool than in a hot greenhouse. Similar results were obtained by BERNSTEIN and AYERS (1951, 1953) with vegetable crops. Many of the plants in their experiments showed salinity damage only when transferred to hot conditions. EHLIG (1940) working with table grape vines, also reported an all or none effect of salinity, depending on temperature. At 90° F (32.2° C) or below, there were few or no signs of leaf burn, but damage was extensive above 100° F (37.8° C). There are also reports of salinity damage during summer months with little or not damage during the winter season (WALL and HARTMAN, 1942; BROUWER, 1963a). Although such reports showing an all or none effect indicate a genuine climate-salinity interaction, it is not always clear which factor or complex of climatic fators, is involved.

FRANCOIS and GOODIN (1972) reported on the interaction of temperature and salinity in their effect on the germination of sugar beet seeds. At low temperatures salinity had no effect, but germination inhibition, due to salinity, increased between 25 and 40° C. An extreme example of temperature salinity interaction affecting the photosynthesis of onion, which does not adjust osmotically, was described in the previous chapter. These laboratory results are in good agreement with field observations showing lower tolerance of onion plants to salinity under hot, than under cool weather conditions (BERNSTEIN and AYERS, 1953).

As noted above, LUNT et al. (1960a, b) studied the interaction of salt and heat as affecting the growth of a number of horticultural plants. They found that the temperature and salinity factors obeyed Baule's principle without any sign of a special interaction. However, most of the reports appear to indicate a genuine increase of salinity damage under hot, as compared to cool, conditions of growth.

C. Radiation and Salinity

There are very few reports on the combined action of radiation intensity and salinity on plant growth. Such as there are usually show greater effect of salinity

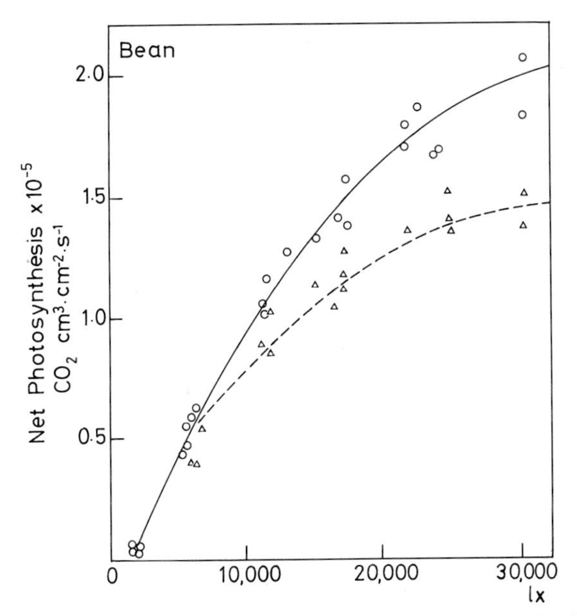

Fig. 1. Effect of light intensity on the net photosynthesis of control and salinized bean plants. Points are from 5 control (o) and 5 salinized (to − 3.4 bars)

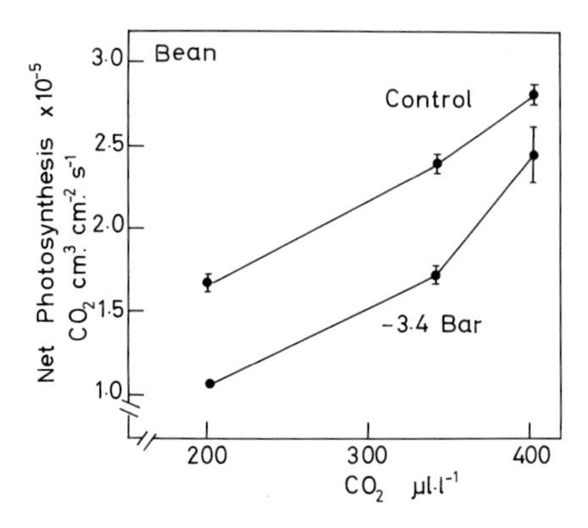

Fig. 2. Effect of carbon dioxide concentration on the net photosynthesis of control and salinized (to -3.4 bars) bean plants at a light intensity of about 26000 lux. At 200 µl/l CO_2 net photosynthesis of salinized leaves was reduced 36% below that of the controls, while at 400 µl/l the reduction was only 13%

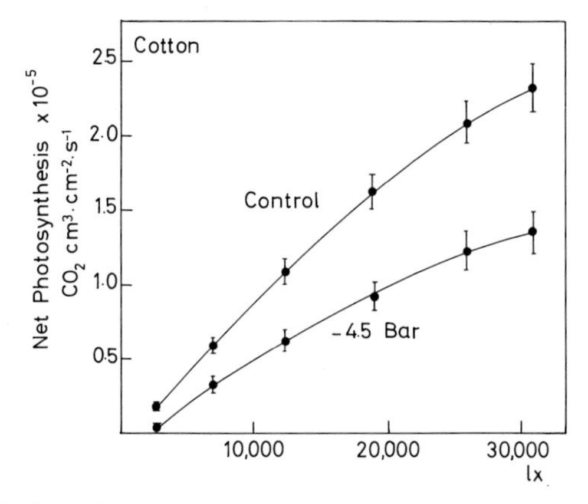

Fig. 3. Effect of light intensity on the net photosynthesis of control and salinized (to -4.5 bars) cotton plants

(i.e. depression of growth) under high than under low light intensitites (BROUWER, 1963b; NIEMAN and POULSEN, 1971).

GALE et al. (1967) studied the effect of salinity on net photosynthesis (P) of bean and cotton plants under different light intensities. Some of their results are presented in Figs. 1 to 4. Results shown in Fig. 1 indicate that in bean plants salinity reduced P only at high (30000 lux) radiation levels and not at all below ~8000 lux. On the other hand approximately the same order of reduction of P by salinity was found in cotton, at all light intensities (Fig. 3). This was interpreted as being due to the varying physiological response to salt of the two species. In beans the primary effect of salinity was apparently an interference with CO_2 diffusion into the leaf, due to closure of stomata. Increasing [CO_2] significantly decreased the effect of salinity at high light intensity (Fig. 2). At low light intensities, when CO_2 is no longer a limiting factor, salt had little or no effect. In cotton,

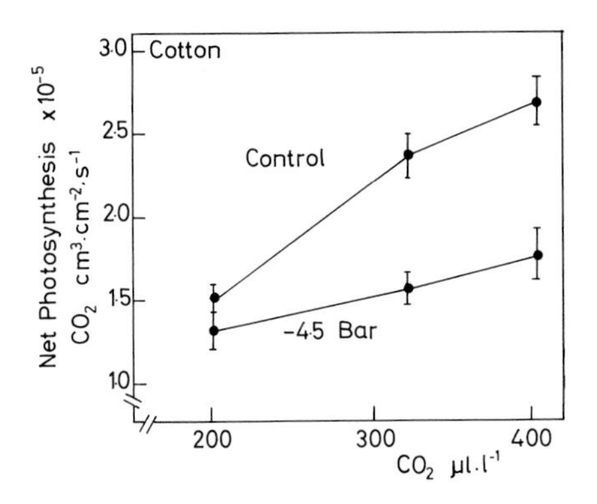

Fig. 4. Effect of carbon dioxide concentration on the net photosynthesis of control and salinized (to −4.5 bars) cotton plants at a light intensity of about 28000 lux. At 200 µl/l CO_2 net photosynthesis of salinized leaves was reduced 13% below that of the controls, while at 400 µl/l the reduction was 34%. [From GALE et al. (1967)]

internal biochemical or photochemical processes appeared to have been damaged and, in contrast to beans, a high concentration of CO_2 could not overcome the effect of salinity at high light intensities (Fig. 4).

To conclude: in plants in which one of the primary results of salinity is a partial closure of stomata, photosynthesis and hence overall plant growth will be reduced relatively more at high than at low levels of radiation. It should of course be remembered that radiation is the major and, in non-advective situations, the only source of energy for transpiration. In those plant species where osmotic adjustment is incomplete (see previous chapter) high levels of radiation will intensify salinity damage, by producing water imbalance.

D. Air Humidity and Salinity

In view of the long period of time during which the "Physiological drought" hypothesis was considered to be valid (see previous chapter) it is surprising that only recently has attention been paid to the influence of atmospheric humidity on salt damage. High levels of humidity result in lowered rates of transpiration and hence could be expected to alleviate the effects of any water imbalance due to salinity. Furthermore the fresh: dry weight ratio is increased under humid conditions. This could reduce the concentration of electrolytes, and such was found to be the case for leaves of bean and cotton plants grown under saline conditions (NIEMEN and POULSEN, 1967). On the other hand LAOUAR et al. (1973) found that in *Sinapis alba*, grown in culture solutions whose osmotic potential was lowered by polyethylene glycol (PEG), osmotic adjustment was faster in a 50% than in a 70% RH atmosphere. However, as a general rule, high humidity has been found to ameliorate growth under conditions of salinity (e.g. PRISCO and O'LEARY, 1973, working with beans and NaCl).

A special case of air humidity-salinity interaction on the growth of the halophyte *Atriplex halimus* was described by GALE et al. (1970). As for many halo-

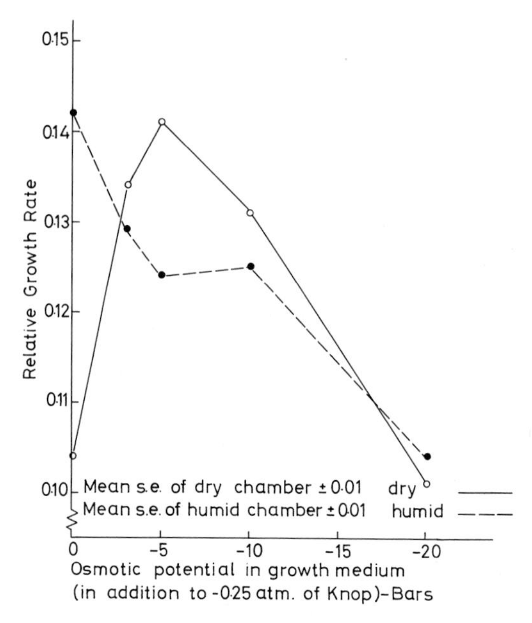

Fig. 5. Effect of salinity on relative growth rate of *A. halimus* plants at two different relative humidities. Mean standard error for both dry and humid chambers is ± 0.01. Values of π are in addition to -0.25 bars of Knop culture solution. Dry Chamber -27% RH, Humid -60% RH, day and night. [After GALE et al. (1970)]

phytes a typical optimum curve is obtained for *A. halimus* when growth is plotted against salt concentration in the growth medium. Optimum growth is usually obtained when the osmotic potential (π) of the culture solution is in the range of -3 to -5 bars NaCl. GALE et al. grew *Atriplex* plants in growth chambers at two levels of humidity 27 or 65% RH. Under the "dry" air conditions the usual optimum curve was obtained. However under the "humid" conditions growth decreased with each increment of salinity between 0 and -20 bars NaCl (Fig. 5).

KAPLAN and GALE (1972) found that the humidity-salt interaction shown in Fig. 5 was due to the plant requiring a relatively small amount of salt in order to adjust its internal π and reduce the rate of transpiration under the "dry" conditions (see previous chapter).

An extensive study of the interrelation of salt and atmospheric humidity has recently been published by HOFFMAN and co-workers. HOFFMAN et al. grew various crop plants to maturity in humidity and temperature controlled, sunlit climate chambers (HOFFMAN and RAWLINS, 1970). Their results showed large species differences but were in close agreement to commonly observed weather-crop-salinity interactions in the field for the different crop species.

Bean plants were shown to react strongly to humidity. In both the controls and plants grown under saline conditions growth increased with increasing humidity. The salinity level at which growth was reduced by 50% was raised by each increase in the level of atmospheric humidity (HOFFMAN and RAWLINS, 1970). The level of salinity which proceduces a 50% reduction in growth is a good criterion for determining true interactions between salt and other environmental factors.

Hoffman and Rawlins (1971) found that when the air humidity was raised from 45 to 90% RH, yield of beet grown in non-saline media increased by 50% and that of radish by 15%. Onion was unaffected. High RH raised the salinity level at which the growth of onion and radish was reduced by 50% but no interaction was found for beet. Both air humidity and salinity had a strong, independent effect on growth of cotton. Even such high humidity did not modify the effect of salinity; the salinity level at which growth was reduced by 50% remained the same (Hoffman et al., 1971).

It appears therefore that air humidity may interact with salinity in their effect on growth of some plant species. Whether or not high humidity will alleviate the effects of salinity will probably depend upon the ability of the plant to adjust internal π and to reduce loss of water thus maintaining its internal water balance.

Although conclusive experimental evidence is lacking, it is probably safe to say that in the final analysis non-edaphic environmental factors may intensify the damage caused by salinity in one or both of two ways: by causing high rates of transpiration or by causing high leaf temperatures. Leaf temperatures and transpiration rates are determined by the interaction of many plant and environmental factors. These include: leaf dimensions, stomatal and internal leaf resistances to water loss, optical properties of the leaves, the radiation environment, wind velocity, air temperature and humidity. The effect of any one of these factors alone varies in degree and direction depending on the values of the other component factors of the leaf energy budget equation (Raschke, 1956; Gates, 1968).

References

Ahi, S. M., Powers, W. L.: Salt tolerance of plants at various temperatures. Plant Physiol. **12**, 767–789 (1938).

Bernstein, L., Ayers, A. D.: Salt tolerance of six varieties of green beans. Am. Soc. Hort. Sci. Proc. **57** 243–248 (1951).

Bernstein, L., Ayers, A. D.: Salt tolerance of five varieties of onions. Am. Soc. Hort. Sci. Proc. **62**, 367–370 (1953).

Brouwer, R.: Some physiological aspects of growth factors in the root medium on growth and dry matter production. Jaarb. Inst. Biol. Scheikundig 11–30 (1963a).

Brouwer, R.: The influence of the suction tension of the nutrient solution on growth transpiration and diffusion pressure deficit of bean leaves *(Phaseolus vulgaris)*. Acta Botan. Neerl. **12**, 248–260 (1963,b).

Ehlig, C. F.: Effects of salinity on four varieties of table grapes grown in sand culture. Am. Soc. Hort. Sci. Proc. **76**, 323–331 (1940).

Francois, L. E., Goodin, J. R.: Interaction of temperature and salinity on sugar beet germination. Agron. J. **64**, 272–273 (1972).

Gale, J., Kohl, H. C., Hagan, R. M.: Changes in the water balance and photosynthesis of onion, bean and cotton plants under saline conditions. Physiol. Plant. **20**, 408–420 (1967).

Gale, J., Naaman, R., Poljakoff-Mayber, A.: Growth of *Atriplex halimus* L. in sodium chloride salinated culture solutions as affected by the relative humidity of the air. Australian J. Biol. Sci. **23**, 947–952 (1970).

Gates, D. M.: Transpiration and leaf temperatures. Ann. Rev. Plant. Physiol. **19**, 211–238 (1968).

Hoffman, G. J., Rawlins, S. L.: Design and performance of sunlit climate chambers. Trans. Am. Soc. Agr. Eng. **13**, 656–60 (1970).

HOFFMAN, G. J., RAWLINS, S. L.: Growth and water potential of root crops as influenced by salinity and relative humidity. Agr. J. **63**, 877–880 (1971).

HOFFMAN, G. J., RAWLINS, S. L., GARBER, M. J., CULLEN, E. M.: Water relations and growth of cotton as influenced by salinity and relative humidity. Agr. J. **63**, 822–826 (1971).

KAPLAN, A., GALE, J.: Effect of sodium chloride salinity on the water balance of *Atriplex halimus*. Australian J. Biol. Sci. **25**, 895–903 (1972).

LAOUAR, M. S., VARTANIAN, N., VIEIRA DE SILVA, M. J.: Effects de l'interaction de l'humidité atmosphérique et due potential osmotique de la solution de culture sur les réactions hydriques du *Sinapis alba* L.: potentiels hydriques et osmotiques dans la plante. C. R. Acad. Sci. Paris Ser. D. **276**, 41–44 (1973).

LUNT, O. R., OERTLI, J. J., KOHL, H. C.: Influence of environmental conditions on the salinity tolerance of several plant species Trans. 7th Intern. Congr. Soil Sc. 560–570 (1960a).

LUNT, O. R., OERTLI, J. J., KOHL, H. C.: Influence of certain environmental conditions on the salinity tolerance of *Chrysanthemum morifolium*. Proc. Am. Soc. Hort. Sci. **75**, 676–87 (1960b).

NIEMAN, R. H., POULSEN, L. L.: Interactive effects of salinity and atmospheric humidity on the growth of bean and cotton plants. Botan. Gaz. **128**, 69–73 (1967).

NIEMAN, R. H., POULSEN, L. L.: Plant growth suppression on saline media: interactions with light. Botan. Gaz. **132**, 14–19 (1971).

PRISCO, J. T., O'LEARY, J. W.: Effects of humidity and cytokinin on growth and water relations of salt-stressed bean plants. Plant and Soil **39**, 263–276 (1973).

Raschke, K.: Über die physikalischen Beziehungen zwischen Wärmeübergangszahl, Strahlung, Austausch, Temperatur und Transpiration eines Blattes. Planta **48**, 200–237 (1956).

TRELEASE, S. F., LIVINGSTON, B. E.: The relation of climate conditions to the salt proportion requirement of plants in solution cultures. Science **59**, 168–172 (1924).

WALL, R. F., HARTMAN, E. L.: Sand culture studies of the effects of various concentrations of added salts upon the composition of tomato plants. Am. Soc. Hort. Sci. Proc. **40**, 460–466 (1942).

General Discussion

A. POLJAKOFF-MAYBER and J. GALE

The response of plants to saline environments has evoked interest mainly from two points of view. One, purely applied, is concerned with the utilization of saline soils, the prevention of secondary salinization and the usage of slightly saline waters in agriculture. The other is concerned with the basic effects of salinity on structure and composition of soils, the interrelationships between salinity of soils and water, and the plant's growth, productivity and physiology.

Although the study of the latter problems may be described as purely basic science, it forms the foundation for an intelligent approach to the solution of the problem of salinity in a way beneficial to mankind. This applied aspect of basic research on salinity seems at times a little academic, since many of the agrotechnical solutions devised to cope with salinity were and are based on trial and error. Nevertheless an intelligent approach to problem solving and a thorough understanding of man's effect on the environment requires this basic information.

The problem of salinity is very widespread. Salt marshes of one type or another, and salt affected areas are found in all parts of the world (CHAPMAN, 1960 and Chapter I; MCGINNIES et al., 1968). Moreover, nowadays, with the spread of intensive methods of agriculture, no country can consider itself immune against the occurrence of secondary salinity. There is also a growing need to bring new areas under cultivation or to reclaim for cultivation areas that were abandoned due to secondary salinity (e.g. reclamation for pasture, MALCOLM 1971, 1972).

The natural saline habitats are distinguished by their special vegetation of halophytes. Apparently no attempt has ever been made to domesticate such plants. Even at present the information on the growth rate, extent of land coverage and nutritive value of such halophytes is very meagre (LACHOVER and TADMOR, 1965; MALCOLM, 1971). Such information is much needed.

Agricultural crops and cultivated plants in general show marked differences in their salt tolerance. There is also a great variability between different varieties of each species. The procedure for determination of crop tolerance level to salinity is rather tedious (CARTER, Chapter II; RICHARDS, 1954). An attempt to establish the level of tolerance, using tissue culture as a model, was unsuccessful (STROGONOV et al., 1970). Isolated roots grown in culture media (Lucern) were much more sensitive to salinity than intact Lucern plants. Callus cultures also were not a suitable model. The question—which type of tissue to take for the culture and when during the year to make the test—remained unresolved. Moreover, callus cultures from the halophyte-*Salicornia*—were as sensitive to salinity as callus cultures from carrots, cabbage, sorghum or tobacco. STROGONOV et al. (1970) conclude that salt tolerance is a characteristic of the intact plant. However the intact plant shows varying tolerance to salinity at different ages and developmental stages.

The extent of damage induced by salinity depends not only on the level of salinity in the soil, or in the water, but also on other external factors, such as soil aeration and climatic conditions (see Chapters II, III, IV, IX, and X). Unfortunately, at present there is not enough information available to explain the interactions between environmental factors. New detailed information is needed on the relative importance of the different parts of the root system (at different periods of growth and at different seasons of the year) in supplying the water requirement of the plant. The solute concentration and the ionic composition of the soil water may vary in different regions of the root system (Chapter IV); some roots may even reach the ground water (this is especially relevant for perennials, both cultivated and wild). There is no information available as to whether coordination exists between roots in different soil regions or how such coordination may be achieved; although it is known that excessive root development may take place in more favorable soil regions. Sometimes even a single root may be quite capable of supplying the water requirement of an entire plant (MARTIN and CLEMENTS, 1939; KRAMER, 1949; EPSTEIN, 1972).

Considerable information exists on the ionic balance of soils under cultivation, on the effect of various types of salinity, on the structure and agricultural properties of soils and on the ionic composition and properties of waters used either for irrigation or for animal and human consumption (Chapters II, III, and IV; HADAS et al., 1973). Much less is known about the properties of soil and water in natural saline habitats. The trend of modern research is to consider the ecosystem as a whole and to investigate the soil—water—plant—complex and the effect of man on the natural equilibrium of the environment. The disturbed equilibrium is sometimes considered as a failure in cycling of the mineral elements or a failure in hydrological cycling. EPSTEIN (1972) summarizes this approach and also summarizes the meagre information available on the interaction of plants with the environment in natural habitats, mainly from the point of view of mineral nutrition. PECK (Chapter V) outlines the complicated interrelationship between the various factors of the environment and interference by man, on salinization and desalinization of the water resources. He illustrates this with the example of the cutting down of forests, in catchment areas, in Australia.

As related in Chapter IX, in the second half of the 19th century, SCHIMPER introduced the concept of "physiological drought" which dominated ideas concerning the mechanism of salinity damage in plants, for almost half a century (SCHIMPER, 1898). This so called "osmotic theory" was challenged by BERNSTEIN (1961, 1963, 1964) and by others (SLATYER, 1961; GREENWAY, 1962; LAGERWERFF, 1969). BERNSTEIN has shown that most plants do adapt osmotically on development of salinization, and that the gradient for water flow into the plant is maintained. Concomitantly, resistance of leaves to water loss increases with increasing salinity. This and osmotic adjustment tend to overcome the increased resistance to water uptake in the roots, which is observed in some plants under saline conditions. However, not all plants show complete osmotic adjustment. Even in many of those that do, under conditions of high evaporation demand of the atmosphere, root resistance may become the dominant factor, and thus bring about a decrease of turgor.

Osmotic adjustment is achieved mainly by active salt accumulation in the cells. However, in some plants salt absorption is passive—with the transpiration stream (e.g. NaCl uptake by tomatoes—SLATYER, 1961). Many glycophytes absorb sodium into the roots, but its transfer to the tops is blocked (JACOBY, 1964, 1965; ESHEL and WAISEL, 1965; SOLOVEV, 1967). In such plants osmotic adjustment is attained by breakdown of polysaccharides or release of potassium ions from bound sites within the cells. The sodium "barrier" between roots and tops may be advantageous; for example, GREENWAY (1962) found a correlation between salt tolerance of different varieties of *Hordeum* and their ability to exclude sodium and chloride ions from the tops. An extreme case is that of *Dunaliella parva* where osmotic adaptation is achieved by accumulation of high concentrations of glycerol (see Chapter VIII).

When sodium is not excluded from the tops, which is the case for most halophytes, plants may expend considerable energy for its extrusion, for example by salt glands (see Chapter VII). Even within the cells, energy may be expended to compartmentalize the ions and exclude them from sensitive organs as shown by LORIMER and MILLER (1969), for KCl in mitochondria from corn.

There seems to be no evidence that even the most halophytic plants thrive under highly saline conditions. At best plants may tolerate, strategically avoid, or otherwise cope with salinity, but usually they grow better under conditions of low salinity.

Extensive research work at the cellular metabolic level has been carried out in the USSR and is summarized by STROGONOV et al. (1970) and STROGONOV (1973). STROGONOV assumed that the first plants, in evolution, evolved in the sea, and as the sea is saline, he suggested that the halophytes preceded the glycophytes. He also suggested that intensive studies of the marine flora lead to an understanding of salt tolerance.

STROGONOV et al. (1970) claim that under saline conditions the balance of photosynthetic pigments is upset. In more sensitive plants chlorophylls are destroyed due to salinity, in more tolerant plants the chlorophyll content increases. From indirect evidence they conclude that salinity affects the strength of the forces binding the complex of pigment-protein-lipid, in the chloroplast structure. As a result, structural changes in chloroplasts are induced by salinity. They state that increase in chlorophyll content, in the more tolerant plants, may be due to accumulation of either chlorophyll-a or chlorophyll-b, but in the sensitive plants the decrease in the amount of chlorophyll is due mainly to the destruction of chlorophyll-a, which is considered to be more labile. They found that in the more sensitive plants, oxidized forms of carotenoids accumulate, and this is considered to have a damaging effect. Besides chlorophylls and carotenoids, the anthocyanins are also affected. The accumulation of anthocyanins is considered by them to be a protective adaptation (STROGONOV et al., 1970).

Other metabolic changes, according to STROGONOV et al. (1970) are the accumulation of keto acids and the incomplete utilization, in synthetic processes, of sucrose which is transported from the tops into the roots. Protein synthesis is inhibited and toxic amino acids accumulate (see Fig. 1). Increased synthesis of anthocyanins and of amide is considered to be a detoxication process. Accumulation of sulphoxides is also considered to be one of the damaging effects of salinity (STROGONOV et al., 1970).

According to STROGONOV et al. (1970) the accumulation of ammonia, certain amino acids, diamines, sulfoxides and sulfonic substances, causes formative changes in the tissues, pigmentation of necrotic spots and salt poisoning of the cells. Carbohydrates, organic acids, some amino acids (mainly proline), pigments

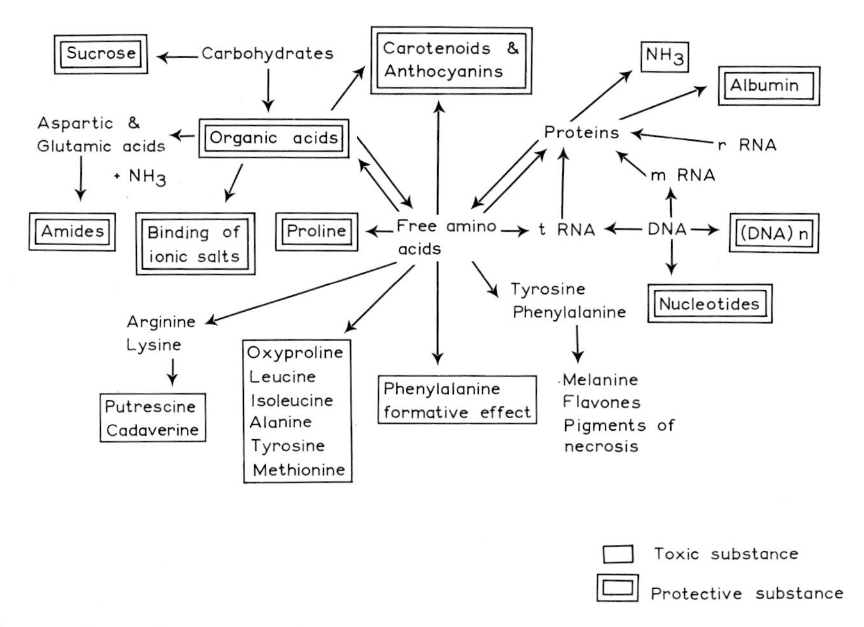

Fig. 1. A scheme of interrelations between various metabolic intermediates, in plants grown in a saline environment. For details see text. [Adapted from STROGONOV et al. (1970)]

such as anthocyanins and carotenoides, nucleic acids and proteins have protective properties. They suggest that survival under saline conditions depends upon the regulation of metabolic processes and the quantitative ratio between the "protective" and the "toxic" intermediates of metabolism (Fig. 1). According to STROGONOV et al. (1970) salinity also induces chromosomal changes, heteroploidy and polyploidy (3 n being considered as the most tolerant state although the evidence for this is rather unconvincing). Most of their conclusions are based on studies of cultivated plants of different degrees of tolerance. The information they present on parallel processes in halophytes, is insufficient.

The ideas suggested by STROGONOV et al. (1970) are interesting. However, it seems to us that they are not supported by sufficient evidence; they were summarized here as some of them may be considered as hypotheses worthy of further investigation.

The biochemical reactions in all metabolic processes are activated by enzymes. The activity of these large protein molecules is due to specifically located active sites which must maintain a certain configuration in order to remain active. The high concentration of ions, accumulating in the cell as a result of osmotic adjustment, may exert an allosteric effect on the enzyme proteins and may affect the structure of membranes, thus changing the natural equilibria of permeability and rates of reactions. The high ionic concentration may also perturb the structure of the solvent—water—and thus also affect the structure of enzymes and membranes (KLOTZ, 1958).

The effects mentioned above, of the high ionic concentration in the cell, are probably responsible for most of the observed changes in the biochemical pathways and enzyme activity induced by salinity (Chapter VIII). These effects may also have much to do with the extremely varied morphological, anatomical and submicroscopic structural modifications occurring in response to salinity (Chapter VI).

Most of the observed changes, metabolic and structural, are probably signs of salinity damage. Only a few of them may be considered to be of a truly adaptive value. Such may be for example:

(a) Increased concentration of the hormone abscissic acid. This induces stomatal closure, thus reducing transpiration and passive uptake of salt. The lowered rate of transpiration also results in an increase of the water content of the tissue which reduces the ionic concentration within the plant. However the partially closed stomata interfere with CO_2 uptake and hence photosynthesis and growth are reduced (Chapter IX).

(b) Increased dark respiration. This supplies energy for pumping salts against electrochemical gradients and facilitates essential compartmentalization. It may also provide energy for increased maintenance—rebuilding of denatured macromolecules. Increased respiration is however at the expense of net CO_2 fixation, which means reduced overall growth (Chapters VIII and IX).

(c) Preference for crassulaceaen acid metabolism versus C_3-carboxylation in CO_2 fixation. This phenomenon, reported for some halophytic plants, results in a much higher ratio of CO_2 fixed per unit water transpired. Its disadvantage is the low capacity of the system for dark CO_2 fixation (Chapter IX).

One might expect to find high salt tolerance or even salt requirement in enzymes isolated from halophytes—similar to what is known for halophilic bacteria—or perhaps stability, such as that conferred by the presence of endogenously generated glycerol on enzymes isolated from *Dunaliella parva* (Chapters VIII and IX). However, there is no indication that this is so, and enzymes isolated from halophytes are as sensitive, *in vitro*, to salinity, as the enzymes isolated from glycophytes. The reason for this may be that many enzymes in the cells of halophytes are not actually exposed to high ionic concentrations due to the large degree of compartmentalization of the ions in the cell. Attempts were made to find out whether such compartmentalization really exists. Studies on the location of Na and Cl ions inside the cells did not give clear cut results (NEEMAN, 1968). The general impression gained was that sodium has a high affinity for membranes, while chloride spreads all over the cell. Antimonate granules, marking the localization of sodium, were found inside nuclei and mitochondria and also in the intercellular spaces near the cell walls, and between the cell wall and the plasmalemma. In the salt glands of *Avicennia* and *Tamarix* the Antimonate precipitate was found mainly in the organelles of the cytoplasm, in the nucleus and nucleolus, and to some extent in the mitochondria (SHIMONY, 1972). Considerable granulation was found between the cuticle and the cell walls of the excreting cells. According to these data, it is possible to accept the compartmentalisation of sodium but it is more difficult to accept that of chloride. In summary it may be said that the damage caused by salinity is apparently at least of a dual nature; an osmotic effect which may be induced experimentally by any type of osmoticum, not necessarily by electrolyte salts; and specific ionic effects, of which there are many different manifestations.

Apparently both the osmotic and specific ion effects operate at the cellular-metabolic level and also at higher levels of organization. At the metabolic cellular level the effect may be on membranes and various biochemical reactions, as summarized by KYLIN and QUATRONO (Chapter VIII) and on the submicroscopic structure (Chapter VI). At the higher level of organization the effect may be on various plant resistances to water movement and gas exchange, translocation, growth and development (Chapters IX and X).

There is a considerable variability in these responses between species and even between varieties, but the biggest contrast is between the salt sensitive (glycophytes) and salt tolerant (halophytes) plants.

The large variations in response of different plant species and varieties to salinity can be related to differences in: (1) Their ability to exclude salt from sensitive tissues, cells and organelles (compartmentalization); (2) the ability to achieve complete osmotic adjustment; (3) the inherent stability of membranes, macromolecules and enzyme systems to a milieu of high ionic concentration; (4) the ability to generate factors stabilizing macromolecules such as kinetin and glycerol and finally; (5) the ability to carry out other adaptive modifications, at the lowest possible cost to overall plant growth.

References

BERNSTEIN, L.: Osmotic adjustment of plants to saline media. I. Steady state. Am. J. Botan. **48**, 909–918 (1961).

BERNSTEIN, L.: Osmotic adjustment of plants to saline media. II. Dynamic phase. Am. J. Botan. **50**, 360–370 (1963).

BERNSTEIN, L.: Salt tolerance of plants. Agricultural Information Bull. No. 283, United States Department of Agriculture (1964).

CHAPMAN, V.J.: Salt marshes and salt deserts of the world. London: Leonard Hill 1960.

EPSTEIN, E.: Mineral nutrition of plants: principles and perspectives. New York: John Wiley and Sons 1972.

ESHEL, Y., WAISEL, Y.: The salt realtions of *Prosopis farcata* (Banks et Sol.) Eig. Israel J. Botan. **14**, 50–51 (1965).

GREENWAY, H.: Plant response to saline substrate. I. Growth and ion uptake of several varieties of *Hordeum vulgare* during and after sodium chloride treatment. Australian J. Biol. Sci. **15**, 16–38 (1962).

HADAS, A., SWARTZENDRUBER, D., RIJTEMA, P.E., FUCHS, M., YARON, B.: Physical aspects of soil water and salts in ecosystems. Ecological Studies IV. Berlin-Heidelberg-New York: Springer 1973.

HALL, J.L., FLOWERS, T.J.: The effect of salt on protein synthesis in the halophyte *Sueda maritima*. Planta **110**, 361–368 (1973).

JACOBY, B.: Function of bean roots and stems in sodium retention. Plant Physiol. **39**, 445–449 (1964).

JACOBY, B.: Sodium retention in excised bean stems. Physiol. Plant. **18**, 730–739 (1965).

KLOTZ, I.M.: Protein hydration and behavior. Science **128**, 815–822 (1958).

KRAMER, P.J.: Plant and soil water relationships. New York: McGraw-Hill 1949.

LACHOVER, D., TADMOR, N.H.: Seasonal fluctuations in the mineral content of saltbush (*Atriplex halimus* L.). Israel J. Agr. Res. **15**, 183–189 (Hebrew) (1965).

LAGERWERFF, J.V.: Osmotic growth inhibition and electrometric salt tolerance evaluation of plants. A review and experimental assessment. Plant and Soil **31**, 77–96 (1969).

LORIMER, G.H., MILLER, R.J.: The osmotic behavior of corn mitochondria. Plant. Physiol. **44**, 839–844 (1969).

MALCOLM, C. V.: Plant collection for pasture improvement in saline and arid environments. Tech. Bull. No. 6. Western Australian Department of Agriculture (1971).

MALCOLM, C. V.: Establishing shrubs in saline environment. Tech. Bull. No. 14. Western Australian Department of Agriculture (1972).

MARTIN, E. V., CLEMENTS, F. E.: Adaptation and origin in the plant world. I. Factors and functions in coastal dunes. Carnegie Institution of Washington Publication, No. 521 (1939).

MCGINNIES, W. G., GOLDMAN, B. J., PAYLORE, P.: Deserts of the World: An appraisal of research into their physical and biological environment. Tucson: University of Arizona Press 1968.

NEEMAN, EMA: Attempts to localize the absorbed sodium and chloride inside root cells. MSc Thesis (in Hebrew) The Hebrew University of Jerusalem (1968).

RICHARDS, L. A. (ED.): Diagnosis and improvement of saline and alkali soils. Agricultural Handbook 60. United States Department of Agriculture (1954).

SCHIMPER, A. F. W.: Pflanzengeographie auf physiologischer Grundlage. 1st ed. Jena: Gustav Fischer 1898.

SHIMONY, C.: The ultrastructure and function of salt secreting glands in *Tamarix aphylla* and *Avicennia marina* (Forssk) Vierh. Ph. D. Thesis, submitted to the Hebrew University of Jerusalem (In Hebrew) (1972).

SLATYER, R. O.: Effects of seven osmotic substrates on the water relations of tomato. Aust. J. Biol. Sci. **14**, 519–540 (1961).

SOLOVEV, V. A.: Ways of regulating a surplus of absorbed ions in plant tissues (Na ions used as examples). Sov. Pl. Physiol. **14**, 1093–1103 (1967).

STROGONOV, B. P.: Metabolism Rastenyi v uslovyakh zasolniya (Metabolism of plants under saline conditions). Timiryazev Lectures XXXIII. USSR Academy of Sciences. Moscow: Nauka 1973.

STROGONOV, B. P., KABANOV, V. V., SHEVJAKOVA, N. I., LAPINA, L. P., KOMIZERKO, E. I., POPOV, B. A., DOSTANOVA, R. KH., PRYKHOD'KO, L. S.: Struktura i Funktziya Kletok rastenii pri Zasolenii (Structure and function of plant cells under salinity). Moscow: Nauka 1970.

WAISEL, Y.: Biology of halophytes. New York: Academic Press 1972.

Plant Index

Subject Index

Physical Aspects of Soil Water and Salts in Ecosystems

Editors: A. Hadas, D. Swartzendruber, P.E. Rijtema, M. Fuchs, B. Yaron

221 figures. 61 Tabellen. XVI, 468 pages. 1973 (Ecological Studies, Vol. 4)

ISBN 3-540-06109-6 Cloth DM 99,–
ISBN 0-387-06109-6 (North America) Cloth $38.60

Contents: Water Status and Flow in Soils: Water Movement in Soils. Energy of Soil Water and Soil-Water Interactions. – Evapotranspiration and Crop-Water Requirements: Evaporation from Soils and Plants. Crop-Water Requirements. – Salinity Control.

These collected research papers were read at a symposium in Rehovot, Israel. Theoretical and practical aspects are included and among the subjects covered are the physical aspects of the movement of water and ions in soil, the interactions of water with soil, evaporation from soil and plants, water requirements of crops, and the management of salinity.

Arid Zone Irrigation

Editors: B. Yaron, E. Danfors, Y. Vaadia

181 figures. X, 434 pages. 1973 (Ecological Studies, Vol. 5)

ISBN 3-540-06206-8 Cloth DM 99,–
ISBN 0-387-06206-8 (North America) Cloth $36.20

Contents: Arid Zone Environment. – Water Resources. Water Transport in Soil-Plant-Atmosphere Continuum. – Chemistry of Irrigated Soils-Theory and Application. – Measurements for Irrigation Design and Control. – Salinity and Irrigation. – Irrigation Technology. – Crop Water Requirement.

This book has been written by a large number of specialists for biologists, agronomists, soil scientists, water engineers, and plant physiologists who want a clear presentation of irrigation fundamentals in arid and semi-arid zones. This synthesis of up-to-date information conveys an understanding of the basic principles governing irrigation technology and helps agronomists to overcome the problem of water shortage in arid zone agriculture.

Springer-Verlag
Berlin
Heidelberg
New York

Prices are subject to change without notice